T0318685

The Exposome

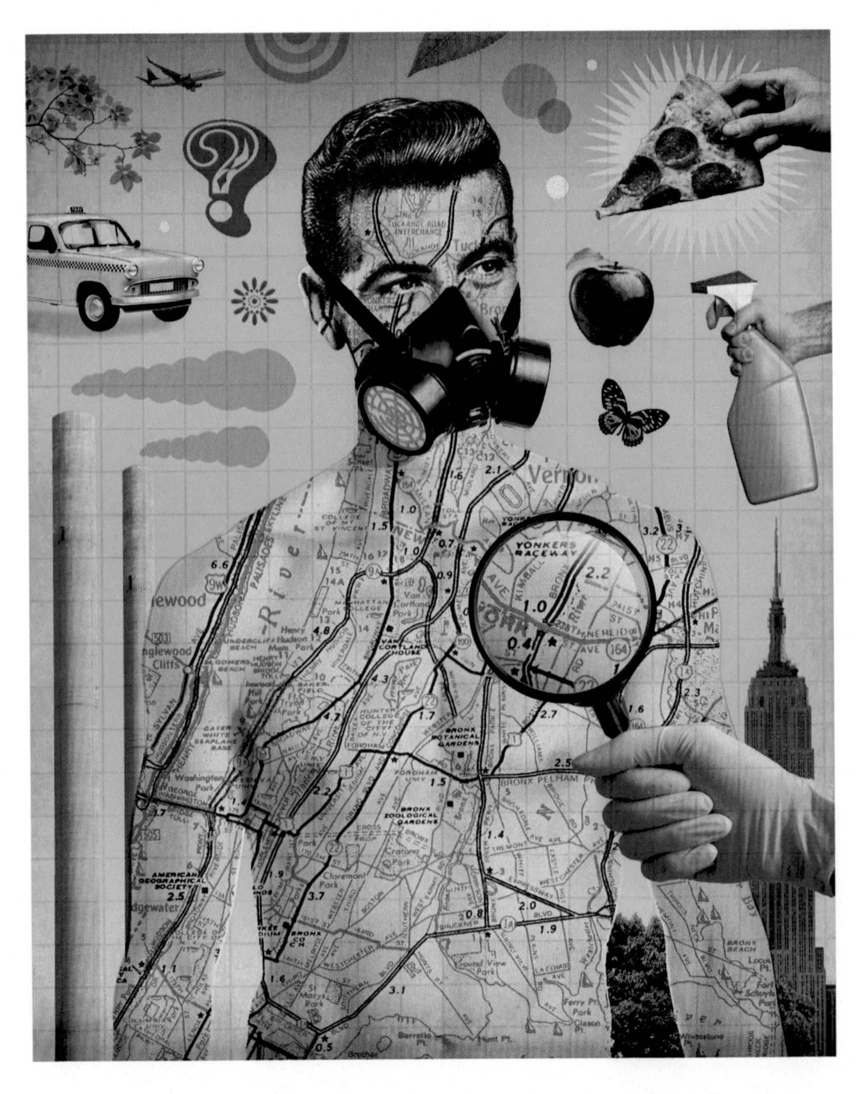

Cover and frontspiece artwork: Michael Waraksa (www.michaelwaraksa.com). Copyright by Michael Waraksa.

The Exposome

A New Paradigm for the Environment and Health

Second Edition

GARY W. MILLER

Vice Dean for Research Strategy and Innovation, Professor of Environmental Health Sciences, Mailman School of Public Health, Columbia University, USA

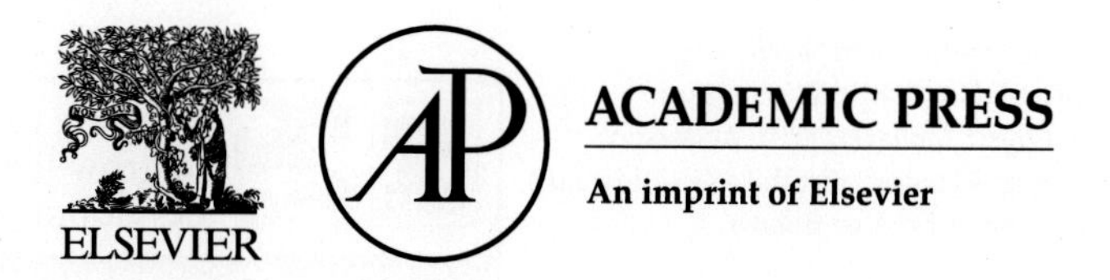

Academic Press is an imprint of Elsevier
125 London Wall, London EC2Y 5AS, United Kingdom
525 B Street, Suite 1650, San Diego, CA 92101, United States
50 Hampshire Street, 5th Floor, Cambridge, MA 02139, United States
The Boulevard, Langford Lane, Kidlington, Oxford OX5 1GB, United Kingdom

British Library Cataloguing-in-Publication Data
A catalogue record for this book is available from the British Library

Library of Congress Cataloging-in-Publication Data
A catalog record for this book is available from the Library of Congress

ISBN: 978-0-12-814079-6

For Information on all Academic Press publications
visit our website at https://www.elsevier.com/books-and-journals

Publisher: Andre Gerhard Wolff
Acquisitions Editor: Kattie Washington
Editorial Project Manager: Kristi Anderson
Senior Project Manager: Kiruthika Govindaraju
Cover Designer: Christian Bilbow

Typeset by MPS Limited, Chennai, India

Dedication

I would like to dedicate this book to my scientific colleagues who have played the roles of supporters, critics, agitators, skeptics, and naysayers (sometimes all at the same time).

Contents

Biography

Gary W. Miller received his Ph.D. in pharmacology and toxicology from the University of Georgia and then completed two postdoctoral fellowships in molecular neuroscience at Emory University and Duke University. He was the founding director of the HERCULES Exposome Research Center, the first NIH-funded center on the topic. His research has been continuously supported by NIEHS since he was a postdoctoral fellow. He has held faculty positions at the University of Texas, Emory University, and Columbia University and has served as an advisor to multiple international projects on the exposome. Dr. Miller wrote the first book on the exposome, *The Exposome: A Primer*, published in early 2014 and served as the Editor-in-Chief of *Toxicological Sciences*, the Society of Toxicology's official journal, from 2013 to 2019. He is currently serving as the Vice Dean for Research Strategy and Innovation and Professor of Environmental Health Sciences at the Mailman School of Public Health at Columbia University in New York where he is building an exposome research infrastructure to better understand the role of the environmental in human health.

Preface to the second edition

Seven years ago I decided to start writing a book about the exposome. The concept was only known within a few academic circles. Times have changed. I believe that the exposome is on the verge of becoming a bona fide scientific discipline. I have heard NIH Director Francis Collins utter the word, as well as the directors of several other NIH institutes. The European Union has invested heavily in projects to pursue the Human Exposome Research Project (which coincides with the timeline I outlined in the first edition of this book). The word exposome is written into French law, and the new Inserm director (France's equivalent to NIH) is supporting efforts on the exposome. What has become clear is that the scientific community is starting to see the value of a systematic and unbiased analysis of the impact of environmental exposures on human health. This is a dramatic shift from a decade ago. The question now is how to do it?

My goal in writing this expanded version of the book was to examine the progress in the field and stimulate thinking about how to chart the path for exposome research over the next decade. While a primer seemed to be the right level of work for the first edition, the exponential growth of the field launched this expanded version and forced a new title for the second edition. The precolon part was clear "The Exposome:" but what about the second half? "A New Paradigm for the Environment and Health"? As the reader will see in Chapter 1, The Exposome: Purpose, Definitions, and Scope, inserting the word paradigm into the title was not taken lightly. This was not about a new paradigm for environmental health, a well-established field, but rather a new paradigm for studying how the environment interacts with our health. More specifically, it is about studying the intersection of our complex environment and health over a lifetime, unrestricted and differently than ways we have studied this in the past. It is about understanding the interface between the environment and human biology. The commentary in the opening chapter addresses the gravity of the word paradigm and my reverence for it.

I have been deliberate in naming names in this book. This is a combination of my desire to recognize the people who have helped shape my exposome thinking but also to provide some personality to the exposome. I am confident that the exposome is going to become a major driver of the study of health and disease, and I believe it is important to note the people who have been involved in the process. I am a terrible note taker, so my reflections are based primarily upon my memory (which is also notoriously spotty). If I have portrayed or neglected any events or people in a way not aligned with reality, I apologize to those I have slighted.

My 16 years as an associate and full professor at Emory could be described as a prolonged postdoctoral fellowship with Dean Jones. He was one of the few colleagues with whom I could discuss scientific questions and not feel as if I was being overly ambitious. Indeed, he often made me think that I was not trying hard enough. His efforts to incorporate high-resolution mass spectrometry into exposome research have been instrumental in driving the field forward and have helped create the field of exposomics.

I would like to thank Kristine Dennis, who was the initial administrator of the HERCULES Exposome Research Center. Her administrative acumen and scientific capacity were exactly what was needed as we established the nascent center. She is now a doctoral student in Dean's lab and I look forward to watching her scientific career advance. Megan Niedzwiecki brought in much needed bioinformatic expertise to my research group as a postdoctoral fellow. She had a willingness to pursue whatever whimsical idea I placed before her and served as a key connector to Dean's bioinformatic team. Vrinda Kalia has been game for all things exposome. While a mere graduate student at Emory and Columbia, she operated at a level much higher and provided a needed public health perspective complemented with an ever-expanding bioinformatic and biochemical skillset. Her contributions were especially critical during the transition to Columbia. I am also indebted to Josh Bradner and Fion Lau for relocating to Columbia and providing a sense of stability to the lab.

I need to thank Doug Walker and Kurt Pennell together. Doug was a graduate student with Kurt Pennell at Tufts University. Kurt and I had worked together when he was at Georgia Tech. Kurt is an environmental engineer who provided the entire Emory contingent important analytical expertise. After molding Doug into a top-notch environmental engineer Kurt sent him to Dean's lab to incorporate exposome-related techniques into his dissertation. Doug was critical in bringing the GC-orbitrap technology into Dean's group, which has helped establish a key workflow into what I now view as exposomics. Somehow, I was fortunate enough to inherit Doug for a brief period as a postdoctoral fellow before he was recruited to Mt. Sinai, and I continue to collaborate with him. I would also like to thank Karan Uppal, a faculty member at Emory who has worked closely with Dean's group. He has been the bioinformatic wizard and has been instrumental in developing the workflow and educating others about it. Even if I did not know or understand the tools needed to solve a particular problem, asking Karan for help became the *de facto* black solution box "Just give it to Karan, he will figure it out."

Few understand the important role Shuzhao Li at Emory University played in making the seemingly impossible possible. The use of mummichog to temporarily sidestep analytical confirmation was essential for us to see a way around the bottleneck. It provided an intellectual shortcut analogous to sequencing the expressed genes à la Craig Venter. The fact that we did not have to wait for all of the features to be identified to drive the biology forward made it all plausible. This was at a time when there was dramatic resistance from the mass spectrometry and chemistry community. Fortunately, that resistance has given way to engagement, insight, and much needed collaborative effort. Even though he was focusing on metabolomics, the approach was apparent to those studying the exposome. After many years at Emory, he has relocated to the Jackson Laboratory to continue his outstanding work.

There are also many colleagues across the world whose names appear on my acknowledgment slide when I give talks, including Steve Rappaport and Martyn Smith at Berkeley, Denis Sarigianis in Greece, Jana Klánová in the Czech Republic, Benedict Warth in Austria, Emma Schymanski in Luxembourg, Shoji Nakayama in Japan, Thomas Hartung in Baltimore, Mark Chadeau-Hyam and Paolo Vineis in London, Martine Vrijhead in Barcelona, Isabella Annesi-Maesano in France, and Marike Kolassa-Gehring in Germany. Roel Vermeulen at Utrecht University has been an important colleague and collaborator, and he, along with Emma Schymanski and Laszlo Barabási at Northeastern Network Science Institute, helped craft what I believe will be an important manuscript for the field. I want to say merci beaucoup to Robert Barouki for hosting my visiting professorship at the University of Paris Descartes in 2018. The timing was perfect. I had made the decision to walk away from the HERCULES Exposome Research Center to move to Columbia, and it was an important time to reflect on how to start anew. Robert remarked that our commonality in thinking about the exposome was grounded in the fact that we are both toxicologists who think about biochemical mechanisms. Essentially, we both viewed the exposome as a way to better understand biology and were not concerned about the potential disruption to existing frameworks that epidemiologists and exposure assessors faced. I would also like to thank Robert Wright, who led Mt. Sinai's CHEAR program and spearheaded the first New York City Exposome Symposium, for his leadership in the field. The Institute for Exposomics that he founded at the Icahn School of Medicine at Mt. Sinai has experienced a meteoric rise. He and the Institute have provided scientific leadership, career advancement, and notable achievement that have added significant clout to the field. The work of his colleague Manish Arora has provided a critical example of how to catalogue exposure history. I also want to acknowledge Kostas Lazaridis at the Mayo Clinic for seeking out a collaboration on the exposome and providing a much needed endorsement of our approach. Our work together exemplifies the power of collaboration. Neither of us knew each other's research area, but through a shared goal of better understanding the role of the environment in human disease we have developed a superb research program. Muredach Reilly, who leads the Irving Institute at Columbia, also recognized the potential and supported the integration of the exposome concept into the realm of clinical and translational research. Richard Mayeux at Columbia has been an early supporter of our approach and I look forward to our exciting collaboration. Along with support of my dean, Linda Friend, and my department chair, Andrea Baccarelli, we are building an important exposome presence at Columbia. Of course, we would not have been able to develop our programs without the support of NIEHS. Former director, Linda Birnbaum, Rick Woychik, and David Balshaw all provided critical support to this burgeoning field. NIEHS hosted an influential workshop in 2015 that helped advance the field.

I also would like to thank Sandra Latourelle from SUNY Plattsburgh. She has taught an exposome course at the undergraduate level and has shared the students' assignments and evaluations with me on more than one occasion, including direct feedback on the first version of this book. Receiving feedback from students who are outside of my own institution helped inform my approach to the new edition of this book. Her efforts also reveal how the exposome can be incorporated into a general biological curriculum.

I also want to acknowledge the members of the Miller laboratory over the past several years, especially the ones who did not directly work on the exposome. Having an innovative and productive research program on environmental drivers of neurodegeneration was a critical grounding for me. Without that scientific progress it would have been difficult to wade out into the uncharted waters of the exposome. I would also like to thank my brother Mark Miller for the excellent editorial expertise. Our careers have converged in an unanticipated manner and it was a pleasure to work together on this project. Lastly, I must acknowledge my wife, Dr. Patti Miller. She has taken on the role of primary reviewer, senior editor, and encourager. The fact that her scientific background is far from exposome research has been instrumental to my ability to develop a narrative that is accessible to a broad range of scientists and general consumers of scientific information. I also want to thank the team at Elsevier. Kattie Washington, Kristi Anderson, and Kiruthika Govindaraju for their patience and assistance. Also, thanks to Erin Hill-Parks and Rhys Griffiths who helped motivate me to write the first edition and to keep a second edition in the back of my mind.

The first version of this book was conceived and written in a matter of months. This version has percolated for much longer. Missed deadlines, unreasonable aspirations, and over confidence in my ability to focus on writing all contributed to the delays, but that additional passing of time allowed for the accumulation of more exposome-related activity. There were also a few professional obligations. Indeed, the six-plus years that spanned the submission of the first and second versions of this book were perfectly filled with my six-year term as editor-in-chief of *Toxicological Sciences*, ending with my relocation from Atlanta to New York City. I could have used many more months to further expand the book and credit the extraordinary work of many other groups, but it is time to get the material out to the next generation of scientists who will be responsible for writing the future chapters that describe a more comprehensive explanation of how the exposome drives human health.

Preface to the first edition

The idea for this book was conceived as the author organized and directed a course on the topic (Genome, Exposome, and Health) at Emory University in 2013. The course was based upon the research of faculty members at Emory University and published work from the primary literature. The students expressed an interest in having access to more background information about the course material. At the time no such material was available. A brief discussion with the editorial staff at Elsevier at a Society of Toxicology meeting led to a book proposal and a rapid turnaround of the text before you.

I would like to thank the scientists who participated in the inaugural course—Dean Jones, Michael Zwick, Matthew Strickland, Jennifer Mulle, Lance Waller, Yang Liu, Yan Sun, Jeremy Sarnat, Dana Barr, Eberhard Voit (Georgia Tech), and Chirag Patel (Stanford). In parallel with the development of the course, Emory University and Georgia Tech developed a Core Center Grant that was funded by the US National Institute of Environmental Health Sciences. The grant, HERCULES: Health and Exposome Research Center at Emory, provides conceptual and technical infrastructure for exposome-related research. A website, humanexposome-project.com, has also been established to provide information to the lay and scientific public.

I am especially thankful to the students who participated in this inaugural course. They were very patient as we developed the course de novo. Their willingness to sign up for a course that included a word not even found in a dictionary or Wikipedia (at that time) is a testament to their inquisitiveness and openness to new ideas. I also want to thank the group of students in the Pharmacology and Toxicology Program at the University of Montana who viewed and critiqued the introductory lecture. Getting feedback from a group of students not associated with our institution helped with the further development of the course.

To faculty members attempting to integrate exposome-based concepts into their curriculum, I believe that you will find the topic to be one that engages and challenges the students (and yourselves). I hope that this introductory text makes it easier to do so. Even though it is in its primordial stage from a practical perspective, the exposome has the potential to play a critical role in our understanding of the environment in human health. Introducing the concept to the upcoming generation of scientists should instill a desire to better understand how the environment impacts health and hopefully inspire them to pursue the challenging questions surrounding the exposome. My desire is that you use the exposome to shamelessly and unabashedly promote the importance of the environment in health and disease.

I do not presume to be an authority on the exposome per se, but rather an environmental health scientist who is exceedingly interested in the concept. I undertook this project as a sole author because I thought it was more important to provide a focused and consistent, albeit idealistic, mindset throughout the book rather than provide an overwhelming, and potentially fractured, compilation on the topic. Given the early stage of the development of the exposome, it would be difficult, if not impossible, to generate such an authoritative tome at this time, even though such a work will be a welcome addition to the field. Certainly, such treatises about the exposome are forthcoming by those that are experts in particular aspects of the exposome, but at this stage, it was my view that an introduction or primer was the most appropriate tack, that is, a course of action meant to minimize opposition to the attainment of a goal. This is necessary because there has been some reluctance, skepticism, and opposition to this topic. When one considers the potential utility of the exposome, it becomes clear that this is, indeed, something that must be pursued. I have chosen to use the first person in those sections where I espouse my views and opinions that may not be consistent with others in the field and I take responsibility for these thoughts. My goal is to engage trainees so that they contemplate and critically analyze the exposome-related concepts and approaches.

CHAPTER ONE

The exposome: purpose, definitions, and scope

Don't undertake a project that is not manifestly important and nearly impossible.

Edwin Land

1.1 Introduction

Let me begin by explaining the purpose of this book. In 2013 I wrote the initial version of this text under the title of *The Exposome: A Primer*. At that time I wanted to provide a brief overview of this exciting and relatively new concept. As I write this second, expanded edition six years later, it is gratifying to see how much progress has occurred in the field. The exposome is a simple concept with extraordinary implications. The exposome can be viewed as the environmental equivalent of the genome as it is an all-encompassing view of the exposures we encounter throughout our lives. This book is for people who are interested in learning more about the exposome and setting the foundation for future study. The text provides an overview of the concept of the exposome and explains how it can be used by students, scientists, healthcare professionals, policy makers, and the general public to better understand the importance of the environment in health. Each chapter provides suggested readings for additional study and discussion questions for further contemplation. That said, this is not solely for the exposome neophyte. The field has been rapidly evolving and I have attempted to capture some of this exciting history. Thus, those who have been pursuing exposome research or are well-versed in the subject should find this expanded version of value as it relates to the history and direction of the field. As I navigate from introductory material to more in-depth analysis, readers will need to be patient with the parts that seem overly simplistic or complex depending on their background. I tried to strike a middle ground. The exposome is

The Exposome.
DOI: https://doi.org/10.1016/B978-0-12-814079-6.00001-8

an empowering concept and should not be left to bespectacled laboratory scientists but rather woven into everyday life. As such, the concepts and language need to be scientifically sound *and* accessible to the general population.

At the risk of sounding pedantic, I must start this book with definitions. Confusion and controversy often erupt when parties disagree on the fundamental meaning of words. In the title of this book *The Exposome: A New Paradigm for the Environment and Health*, there are four words that merit attention: exposome, paradigm, environment, and health. Let us look at them in that order.

1.2 What is the exposome?

The term exposome was first coined by Christopher Wild in a 2005 paper titled *Complementing the genome with an "exposome": the outstanding challenge of environmental exposure measurement in molecular epidemiology* (Wild, 2005). The paper was provocative and demanding. In it, he put forth the first definition of the concept (Fig. 1.1).

Wild, an epidemiologist, expressed concern over what he perceived as a lack of information about our environmental exposures. Specifically, he was concerned with the lack of measures or tools that epidemiologists could use to help identify contributors and causes of human disease. Subsequent publications from Wild described three distinct, but complementary descriptors of the exposome: the internal (the within the body measures), specific external (the immediate local environment, radiation, diet, lifestyle, pollution), and general external (societal, economic, psychological) (Wild, 2012; Wild, Scalbert, & Herceg, 2013). Wild's idea was insightful and timely, but for several years, it did not receive as much attention in the biomedical research community as one may have predicted and hoped. There are numerous explanations for this, but one possible reason is that the paper focused on the exposome as an issue

Original (Wild)

All exposures from conception onward, including those from lifestyle, diet, and the environment

Figure 1.1 ***Exposome definition as originally proposed by Christopher Wild.*** As noted in the text, Wild coined to term to represent the totality of exposures.

relevant for only exposure assessment and epidemiology. Of course, a measurable exposome would be immensely beneficial to these fields, but the potential impact of being able to measure the exposome goes far beyond exposure assessment and epidemiology. All areas of health and biomedical science stand to gain from a better appreciation of environmental forces. Indeed, the exposome has the potential to be even more influential than the genome.

Dean Jones, with whom I worked for over 15 years while at Emory University, and I proposed a slightly modified version of the exposome definition in early 2014 (Miller & Jones, 2014), which I more fully explained in the first edition of this book. We had found that as we interacted with our colleagues, the original exposome definition and interpretation did not mesh well with the U.S. biomedical enterprise. This was partly based on the ability of people to dismiss the concept due to its seemingly impossible goal of measuring the totality of exposures, and partly due to the disconnect with biological sciences. Many of these ideas are captured in Wild's papers, but they were lacking in the prototypical dictionary or textbook definition. Our proposed 2013 definition is in Fig. 1.2.

The notable differences between the original definition and ours are as follows. (1) *The cumulative measure.* That is, the components that make up the exposome are measurable items. If there is no current (or in development) means of measuring the component, then it cannot be part of the entity. It is also an accumulation of the exposures throughout life. (2) *Associated biological responses.* The way our bodies respond to the various external forces is part of what makes up the exposome. The differential response seen among individuals may be as important as the exposure itself.

Expanded (Miller and Jones)

The cumulative measure of environmental influences and associated biological responses throughout the lifespan including exposures from the environment, diet, behavior, and endogenous processes

Figure 1.2 ***An expanded definition of the exposome.*** Here, a new definition by Gary Miller and Dean Jones is provided that focuses on the measurable qualities of the exposome and emphasizes the importance of the body's response to the exposures that occur through one's lifetime. The exposome is a combination of the complex exposures and the complex responses to said exposures. The biological manifestations and responses are arguably as important as the exposures themselves, as they represent the summation and integration of the exposures that occur within the context of our genetic mosaic.

For example, an epigenetic alteration in response to an environmental trigger mediates the ultimate outcome of that exposure. The alteration in DNA one measures at any given time results from the combination of damage coming from exogenous and endogenous sources, as well as the inherent DNA repair mechanisms. When exposures change our biology, *ipso facto*, there must be biological evidence of such. These exposures invariably leave some type of molecular fingerprint. These clues are critical to reconstructing the exposure history. We must include the body's response to the environmental insults in the definition of the exposome. (3) *Explicit insertion of the word behavior.* Specifically, the word is used in a broad context here to capture behaviors that are self-initiated and those that are exerted upon us from outside of our bodies. Considered broadly, one can think of human behavior. What are the behaviors that humans engage in that may influence health? These would include many of the social determinants of health, stressors that result from relationships and habitats, risky behaviors, and positive activities, such as exercise or general physical activity. (4) *The endogenous processes.* Our bodies are an ongoing biochemical experiment. Thousands of biochemical reactions are occurring at any moment, from the breakdown of nutrients to the creation of cells and tissues. These reactions yield many byproducts that can impact our health. For example, such reactions not only generate free radicals that can damage macromolecules but also act as important signaling molecules. In addition, the microbes that live within us and on us can metabolize molecules in our gut, on our skin, or in other organ systems. It should also be noted this expanded definition does not distinguish between an internal or external exposome or the concept of the eco-exposome (see Chapter 3, Nurturing Science). The exposome is by definition all-encompassing and includes elements internal and external to our body. Thus, while this expanded definition provides more detail, it is more inclusive and unifies several current schools of thought about what the exposome is (Fig. 1.3).

Our proposed definition sets the stage for a more comprehensive evaluation of the exposome, and it incorporates some key concepts that no longer need to be inferred from the original definition. We must be reminded that the exposome is massive and will resist any attempt to simplify or divide its character. The following chapters will attempt to expound upon the definition by providing information about the various approaches and models that can be used to develop the structural framework of the exposome.

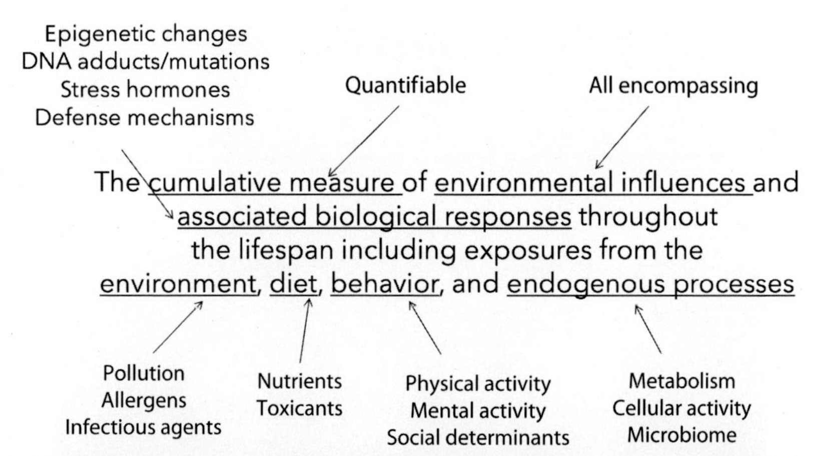

Figure 1.3 ***Anatomy of an updated definition of the exposome.*** The expanded definition proposed by Miller and Jones builds upon the foundation provided by Wild. The key differences are highlighted here. The exposome is a measurable entity that includes a suite of measures. The environmental influences are all encompassing and accumulate throughout life. The associated biological responses, that is the results and responses to those exposures over time, are a key part of the exposome. Exposures to environmental factors, such as pollution, allergens, and infectious agents, as well as the beneficial nutrients and potential hazardous components of our diets, and the concept of behavior. By behavior, we mean our activities (exercise, occupation, hobbies), our mental efforts (education, ongoing learning, engagement in intellectually stimulating activities), and social factors. The social factors are often not considered within the context of environmental health, but the stresses exerted by economic strains, community issues, violence, and our home and social life can have major effects on our health and should be included in the exposome. Lastly is the inclusion of the endogenous processes. The processing and metabolism of chemicals necessary for the function of cells and organs can generate harmful molecular species or be involved in various repair pathways. Our microbiome is a great example of these processes that can impact our health.

1.3 What is a paradigm?

Ever since reading Thomas Kuhn's seminal work on the nature of scientific discovery as an undergraduate, I have been very careful when using the word paradigm. First published in 1962, *The Structure of Scientific Revolutions* (Kuhn, 1962) is one of the most important books on the history and philosophy of science (see Fig. 1.4). In addition to providing insight into how science advances, it popularized the word paradigm and the term paradigm shift. Indeed, I have worked hard to avoid using the word/term as a way of protecting the value of its meaning. Kuhn used

Figure 1.4 The Structure of Scientific Revolutions. The realization that one may have too many copies of a particular book, albeit an important one.

the word in several different contexts, and there has been considerable debate over the precise meaning of the word. I believe it should be reserved for those constructs that truly transform fields. As noted in Chapter 2, Genes, Genomes, and Genomics:Advances and Limitations, the discovery of the structure of DNA serves as an exemplary example, but when the term is used more than once in a conversation or at a particular scientific or business meeting, it is generally safe to say that it is not being used in the manner in which Kuhn intended.

The challenge of explaining the word paradigm stems from the scientist philosopher who popularized the term himself. In *The Structure of Scientific Revolutions*, Kuhn used the word paradigm in several dozen contexts (as we will see, this sounds vaguely familiar). This was the topic of rather heated debates decades ago and even lead Kuhn to add a postscript to the 1969 version of Structure (Kuhn & Hacking, 2012).

In that postscript, Kuhn attempts to summarize his multiple uses of the word paradigm with this:

> *...it stands for the entire constellation of beliefs, values, techniques, and so on shared by the members of a community.*

The entire constellation of beliefs, values, and techniques beautifully describes the ethos of a field. Does the field of the exposome have a constellation of beliefs, values, and techniques? As outlined in later chapters, we certainly have a good collection of techniques, but do we have beliefs and values? For the most part, we have manuscripts that describe exposome research and major research hubs that support exposome research, but I am not aware of any collective beliefs and values, at least not in any clearly articulated or formal way. This was one of my goals when I wrote *The Exposome: A Primer*, and it has been an overarching goal of the current text. This is not to say that I am going to dictate what these beliefs and values are; rather I will discuss the current state and provide a blueprint for how the exposome field can achieve this. This theme is woven into the entire text and will be the major focus of Chapter 10, The Exposome in the Future.

When I state that the exposome is poised to take on the paradigm label, I do not take it lightly and still have trepidations when I see it used in the title of this book. While the exposome concept and its methods may have not yet achieved this stature, I do believe that the current conditions are indicative of the state that directly precedes a paradigmatic shift within a particular discipline. Specifically, fields approaching paradigmatic shifts often find themselves in states of crisis. There is inner turmoil in a field because existing approaches are inadequate for the current or emerging questions. Our inability to address the health effects of exposures that occur as mixtures, that is, essentially all of them, is a case in point.

Kuhn goes on to note

> *...it denotes one sort of element in that constellation, the concrete puzzle-solutions which, employed as models or examples, can replace explicit rules as a* basis for the solution of the remaining puzzles *in normal science.*

To reiterate, the concrete puzzle-solutions that serve as models to solve the remaining puzzles in that domain are a key feature. Even in his attempt to consolidate his definition of the term, he still provided multiple and incredibly meaningful definitions. Those that I quote are most suitable for the exposome, and these puzzle-solutions provide specific examples of how a problem is solved in that particular field. What are these concrete puzzle-solutions for the exposome? Do they exist? Is it the pioneering environmental-wide association study (EWAS) published by Patel, Bhattacharya, and Butte (2010)? Is it the birth cohort approach used by Vrijheid et al. (2014)? Is it the metabolome-wide association approach used in the study of disease? Is it the robotic high throughput Tox21 initiative? I believe these studies and projects represent the types of approaches that could serve as exemplars for future studies, but none provide the singular example, and all have their limitations. Work that relies on National Health and Nutrition Examination Survey (NHANES) data is limited to those 200 to 400 chemicals that are measured under that particular program. Work from untargeted high-resolution mass spectrometry may cover chemical and biological space, but the level of uncertainty and unknowns is exasperating. Studies conducted in cell culture or on organs on a chip cannot capture human complexity. Perhaps, there could there be a combinatorial approach of all of the above that can get us closer to the solution.

I am often asked by colleagues about specific examples (experimental design, methods, procedures, etc.) of how one would do exposome research or conduct an exposome-scale study. Even though I have not been able to point to the ideal study, the fact that colleagues are asking for what sounds like the concrete puzzle-solutions Kuhn endorsed makes me think that we are heading in the right direction. Although I agree that it is important to have such examples, I also appreciate that if it were easy to conduct a study that exemplifies the exposome challenge, it would not merit the development of a new field. We have been and will continue to develop the exemplar studies, but there will likely be several useful examples and it will take time to accumulate the representative case-studies that make up the exposome paradigm. I believe these puzzle-solutions will evolve rapidly over the next few years.

There will be times when the approaches used to study the exposome will appear to be incommensurate with the normal science of other fields, such as epidemiology, toxicology, and exposure science. This is not necessarily a bad thing. It may serve as evidence of the paradigm-preceding crisis events. It may be more comfortable to revert back to the usual science or techniques that have been the standard operating procedures for decades. For the sake of human health, I hope enough scientists accept this discomfort and push forward with the exposome. It will upend many routine ways of doing things and run counter to how people have spent their careers. Ultimately, the approaches outlined in this book will transform the way to study and understand how the environment impacts human health and, in doing so, forever change the way we study the concepts of health and disease. Does this suggest that I was premature in my use of the word paradigm in the title? I do not believe so, but to answer this we must revisit Kuhn.

When one considers a paradigm shift within a field, it is important to consider the circumstances that led to the shift. Essentially, what Kuhn suggests is that fields are approaching a state of crisis in the build-up to a paradigm shift. The recognition that there are unsolvable problems primes the field for new solutions. So what is the emerging crisis facing the field? This requires us to define the field of which we speak. This is addressed in more detail in Chapter 4, The Environment: the Good, the Bad, and the Ugly, but here it is safe to say the field contains those collections of disciplines that are trying to identify environmental contributors to disease. This includes not only environmental health sciences, exposure science, epidemiology, and toxicology, but also all biomedical research that is attempting to understand disease causation. Indeed, it is the recognition of the inability of genetics and heritability to explain the majority of disease variation that has put more attention on nongenetic causation. Thus, the disease specialist contacts experts on environmental factors, and they are offered a list of chemicals that can be measured, information on pollution levels and weather patterns, tools to obtain dietary information, and occupational exposure matrices. After which, the disease specialist realizes that they cannot go back and collect much of this information, that they do not know what chemicals should be assayed from the á la carte menu, and that they cannot afford the thousands of dollars per subject to obtain this information, which leads them to start looking at whole genome sequencing. This pattern has repeated itself over and over in the past few decades.

Those studying the environment have failed to provide tools that are readily adopted by nonspecialists. This point cannot be stressed enough. I believe that this failure, in the face of ever-improving genetic tools, has brought us to the precipice of obsolescence. Occupational medicine residencies have been closing, and physicians receive minimal training about environmental factors in human disease. The environment has been nearly written out of the medical establishment. Even the field of public health with its emphasis on prevention and population science has downgraded environmental health in its educational accreditation requirements.

This is myopic and wrong. Yes, doctors are aware of risk factors for diseases. Smoking, inactivity, and poor diet are major contributors to many complex diseases. Air pollution is a major contributor to asthma and other cardiovascular disorders. Yet, precision medicine is all about how one's genetics dictates their disease probability and treatment response. Once you get past genetic risk factors, *all other risk factors are environmental*, that is, components of the exposome. They represent external forces that are under a varying degree of control of the individual. Eating and activity patterns are a combination of personal behaviors, such as dietary preferences, and externalities, such as access to quality food. Water and air quality tend to be controlled by policy and government decisions. Using a holistic exposome framework may allow a systematic evaluation of these external forces.

The exposome will undoubtedly transform environmental health sciences, but that alone does not make it paradigm-worthy. Buck Louis, Smarr, and Patel (2017) wrote a commentary in which they use the term exposome paradigm and lay out a vision for research in the context of epidemiology. The authors noted that their vision was more prophetic than actionable given the need for improved technologies. At the same time the article was caged in an epidemiological framework, which is important, but the exposome is more than epidemiology. It is as fundamental to biology as is genetics. As outlined in Chapter 10, The Exposome in the Future, the future work on the exposome will serve as the evidence for whether or not the word exposome can justify its presence in the same sentence as paradigm.

1.4 What is the environment?

Defining the environment is a major challenge within an introductory chapter; thus all of Chapter 4, The Environment: the Good, the Bad, and the Ugly, is dedicated to this task. I have always taken an expansive view of our environment, as in the nature/nurture, gene/environment contexts, but this is not how most of science views it. What I have observed is that the environment in a health context is much of what resides in departments of environmental health science- classical occupational exposures, air pollution, water pollution, radiation, industrial chemical exposures, and chemicals in consumer goods. Often, this cluster of exposures is considered to be inadvertent; that is, the individual does not intend to be exposed to these items. However, it is very difficult to infer the health impact of unintentional exposures in the absence of knowledge about the more intentional exposures. The food we ingest, the supplements we take, and the medicines we consume (pharmaceutical or recreational) can have dramatic effects on our health, yet these exposures are placed within other health and medical domains that can make it impossible to integrate the biological effects of the combined exposures. The recognition of the health effects of climate change adds another layer of complexity, but it does not appear that climate change is a new type of exposure, but rather an accelerator or amplifier of many existing exposures. Temperatures, humidity, and disease vectors typically have not been major areas of study in environmental health, but with more dramatic shifts in the global balance of these factors the more obvious it is that they are indeed part of field. Thus while they are being incorporated into environmental health science research and training, they should and will be part of exposome research. For the sake of introduction, the traditional definition of the environment referring to the conditions and surroundings in which a person lives should suffice, but we must expand it to a population level. Chapter 4, The Environment: the Good, the Bad, and the Ugly, explores this in more detail.

1.5 What is health?

For over 70 years the World Health Organization's definition of health has been the "go to" version.

State of complete physical, mental, and social well being, and not merely the absence of disease or infirmity.

I suspect it has been cited tens of thousands of times since its inception. Unfortunately, what it describes is something that appears to be impossible to attain or retain. People with an inherited chronic disorder are by definition unhealthy regardless of their actual health state and their discipline in following a treatment schedule. When such a definition proports an unattainable state, it becomes rather meaningless from an operational perspective.

We need a better conceptualization of health. I believe that one of the more important concepts surrounding health is the ability to respond to challenges. This is often referred to as resilience or robustness. We are not static creatures. Our health status has much to do with how we respond to externalities, which for the sake of the current discussion is represented by our exposome. But I am not sure if resilience or robustness is sufficient. I think a healthy manifestation of the concept of antifragility may be better. The neologism antifragile comes from Nassim Taleb (Taleb, 2012), who is best known for his work *The Black Swan* (Taleb, 2007). *The Black Swan* examined the exceedingly rare events that occur in economics and other business matters that inexplicably occur with seeming regularity. There is a delicious irony in the reoccurring rare events that always seem to take people by surprise, which are referred to as black swans. In his book *Antifragile* (Taleb, 2012), he discusses systems and entities that gain from disorder. He coined the word because in modern language there is not an antonym for fragile (if you disagree, read his book and you will be convinced). While he did not conceptualize this around health, it is a very attractive construct for health. Our bodies do not merely resist exposure to a virus or resist the physical challenges that occur after exercise. Our bodies actually gain from many of the challenges- for example, immunological memory. Being robust just means resisting change, whereas positive adaptation is a more critical aspect of health. If every stressful event to which you are subjected made you stronger, your health status would logically improve. Although on the surface it appears to violate the second law of thermodynamics, it would be an ideal, if not idealistic, component of a revised definition of health. I am not ready to propose a new definition of health but believe that a new definition should consider the ability to adapt to changes in one's exposome in a positive way as a key component. Health is also dynamic, and we should strive toward an operational definition that allows one to integrate the key factors involved in health improvement and support activities that drive the vector in the direction of a higher state of health. This is explained more in Chapter 9, The Exposome in the Community, where I discuss the modifiable exposome.

Now that we have addressed the definitions of those four words, let us go back and examine the derivations of the first word we explored—the exposome.

1.6 Balkanization of the exposome

Wild's early description of the internal, external, and general exposome was the initial step toward derivatization of the exposome term.

I do not disagree with this conceptual framework, but it introduced the idea that there were different kinds of exposomes. Considering that the word was not in our vernacular until 2005 (and in reality, still is not for the majority of science communities and the general public), it is dangerous to fracture its meaning at this early stage. We are trying to convince the funders, regulators, and policy makers that the environment matters, and the exposome is the unifying concept that helps navigate the complexity. It is okay, and actually preferable, for the exposome to be a monolithic exposome.

The Balkan region is located in southern Europe in the areas that were part of the Ottoman Empire, the Austro-Hungarian Empire, and the former Yugoslavia. To balkanize is to divide (a region or body) into smaller mutually hostile states or groups (Oxford English Dictionary). Thus the balkanization of a geographical region describes the fracturing of nation states that commonly portends ongoing conflicts. I fear the exposome could become balkanized (Fig. 1.5).

When we bastardize scientific terms, we risk undermining the underlying science. We should emphasize the singular concept of the exposome and then explain the context in which one is studying it. Thus rather than say eco-exposome, one can define the exposome and then explain

Food exposome
Eco-exposome
Drinking water exposome
Tooth exposome
Placental exposome
Occupational exposome
In utero exposome
Fetal exposome
Urban exposome
Skin aging exposome
Skin exposome
Lung exposome
Socio-exposome
Blood exposome
Endogenous exposome
Public health exposome
Aluminum exposome
Pregnancy exposome
External exposome
Internal exposome
General exposome
In hive pesticide exposome
Pharmaceutical exposome
Plasma exposome
Saliva exposome
Amphibian exposome
Urban lead exposome
Disinfectant by-product exposome
Mitochondrial exposome,
Polar bear exposome

Figure 1.5 ***Balkanization of the exposome.*** A running list of the exposome variants used in the literature. Although it is positive that so many authors are pursuing the exposome, there is a danger in over derivatization. A unified exposome empire is preferable at this time.

how he or she is exploring the ecological influences to the exposome. Rather than the urban exposome, it is the study of the exposome in large urban centers. It is still the exposome. I remain concerned about the use of the word totality in the context of the exposome because of the ease by which one can dismiss a concept that is impossible to define as a practical matter. I have warmed to the occasional use of word totality over the past few years, as it can take quite a bit of time to explain the scope of the exposome.

Discussions of the definitions of the exposome can and do incite some scientific angst. Of course, I prefer the working definition that Jones and I proposed, but that is because it is an operational definition for the type of research we were and are conducting on the exposome. It was fit for purpose, but I fully appreciate that this is viewed from my own scientific perspective as a biomedically based investigator who has been supported by the National Institutes of Health throughout my career. David Balshaw from NIEHS often presents a compilation of definitions when he gives talks and routinely dodges questions about the preferred definition, taking the common apolitical role of a diplomat. These definitions include versions provided by. Christopher Wild, the National Research Council, Germaine Buck-Lewis, Steve Rappaport, and the late Dr. Paul Lioy, and the one that Dean Jones and I put forth (Fig. 1.2).

It is not only derivations of the word exposome that are causing confusion. It is also the use/overuse of the -ome suffix. The first -ome was the genome. Coined in 1920 by Hans Winkler, the -ome suffix is now used to describe the "compilation of all contributing members or the totality of a particular biological domain." Thus, when one considers all of the exposures we face, merging those exposures with the -ome yields "the compilation of all contributing members of our exposures" or "the totality of exposures."

Exposures + -ome = exposome

I routinely cite the riduculome in my talks to poke fun at the explosion of omes, but it is a rather serious matter. Nomenclature within and among fields is important. As we try to merge data across different domains, we are learning more about the importance of ontologies to allow scientists to share information. If we cannot agree on what something is, how do we exchange data? I will discuss some of the other complementary, and at times, silly omes in Chapter 5, Measuring Exposures and their Impacts: Practical and Analytical, but let us revisit what I believe is one of the two most important omes.

Our genome is the blueprint of life. However, as we incorporate genomic information into our decision-making for healthcare and health policy, it is becoming increasingly clear that a genome-centric approach to healthcare is incomplete. This is not to say that knowing one's genetic background is not useful. Indeed, it can be very informative for many situations, but knowing one's genetics is only one part of the equation (more on this in Chapter 2, Genes, Genomes, and Genomics: Advances and Limitations).

Original: phenotype = genotype × environment or $P = G \times E$
Improved: phenotype = genotype × exposome or $P = G \times E$

This concept was introduced to me many years ago in middle school biology, but it was Chirag Patel's guest lecture to my exposome class at Emory University several years that illustrated this truism in the context of the exposome. The E had previously been the environment, but in reality the *exposome is the better definition of E*. Ideally, we define phenotype (P) of being the product of our genotype (G) and exposome profile (E). We have conceptualized genotype, but a parallel exposome-type has not been defined (see Rattray et al., 2018 for the introduction of the term exposotype, which captures much of this topic). As discussed in Chapter 2, Genes, Genomes, and Genomics: Advances and Limitations, our understanding of the G symbol of this equation has benefited from a series of extraordinary advances to the point that we can obtain an impressive level of information about one's genetic background. Unfortunately, a systematic way of capturing an analogous level of information on the environmental influences on our health is lacking.

The genome encodes a plan for what should occur, but the exposome can stifle or amplify the encoded plan. A risk allele for a disease changes probabilities for what will occur when confronted with *the environment*. As conceived, the exposome can be considered a mechanism by which we can quantify nurture. It can be argued that the exposome is unable to capture all of nurture, and this has been used as a reason to dismiss the concept. Every few weeks, the mysteries of the human genome continue to reveal themselves. Between the time of Watson and Crick's discovery and the sequencing of the human genome, one could have used a similar argument. Without the entire genome (including the noncoding elements), the efforts would be fruitless. As we continue to learn, the sequence of the genome merely scratched the surface of genomic knowledge. Geneticists work under a framework that assumes that all mysteries of the genome will eventually be solved, and that confidence allows them to identify the genetic contributors to diseases, traits, and behaviors without apology. Why is it that those interested in broad environmental influences must have their sciences perfected before serious energy and resources can be devoted to the task? Inheritance studies clearly show a major component, namely, the majority, that is not explained by inheritance. Geneticists have known that there are nongenetic contributors, but they focus on the genes. Geneticists are right to ignore the nongenetic effects, because they are geneticists. However, NIH or the European Commission should not operate within a genome-centric bubble. We expect GWAS-level analysis (or better) of disease, but we do not demand an approach that captures the remaining portion. This is unacceptable and represents one of the major drivers of exposome research.

A contrarian view of the exposome as paradigm would be that it is not the exposome itself that represents the new paradigm, but rather it is omic-scale biology that represents the paradigm. If true, the exposome would qualify merely as a subdiscipline. Undoubtedly, omic-scale analysis has transformed the way biology is conducted--moving from studying single genes or proteins and single effects to studying the complication of genes or proteins and complex outcomes. Yet when one looks at omic-scale biology over the past 20 years, the environment is not a member of the club. Environmental health has adopted many approaches of omic-scale biology, such as transcriptomics and proteomics, but has not created its own. It is my view that environmental health has trailed other fields primarily because until now there has not been an omic-scale toolkit for environmental factors.

1.7 Darwin would be proud

Ever since humans (*Homo sapiens*) walked the planet, there has been an awareness of the influence of environmental factors on one's health. Even the Neanderthal (*Homo neanderthalensis*) could identify the predator as an external threat. Indeed, in those prehistoric times, it may have been easier to recognize the exogenous forces acting upon our species than the endogenous factors. No textbooks were needed to convince the Neanderthal that getting hit in the head by a large rock was a bad thing. Nor was it necessary to be told to run when being pursued by a saber-toothed cat. Poisonous plants were avoided after observing the demise of one's associate after consuming the plant. Recognition of the adverse effects of external forces on our health has been intuitively obvious for millennia. Move forward to more modern times. References to the manipulation of external factors abound, while the understanding of internal forces was somewhat limited until the past few centuries. While the practice of Mithridization (more on this in the toxicology section of Chapter 4, The Environment: the Good, the Bad, and the Ugly) likely induces some iatrogenic effects, these practitioners were concerned about an impending environmental insult and took action using a carefully prepared exogenous preparation. Something as simple as how blood flowed throughout the body was not understood until the 17th century, when William Harvey proposed the model of circulation that revolutionized physiology and medicine. Prior to this, humors (black bile, yellow bile, phlegm, and blood), first introduced by Hippocrates and propagated through Galen, ruled the day. These four substances were thought to control our physiology. One could monitor the effects of external forces by examining the humors. Even literature documents the importance of chemical manipulation of one's environment. From Chaucer's *Canterbury Tales* to Shakespeare's "Romeo and Juliet," characters sought the counsel (and drugs) of their local apothecary ("though in this town there is no apothecary, I will teach you about herbs myself" to "O thru apothecary, thy drugs are quick"). They clearly recognized the power of exogenous substances to influence health (Shakespeare, Orgel, & Braunmuller, 2002).

Centuries later, the connection between exogenous germs and disease was revealed. In Vienna, Semmelweis's recognition of the need of surgeons and doctors to wash their hands between autopsies and the delivery of babies, which was initially met with a great deal of skepticism, was ultimately confirmed by Pasteur and led to a revolution in the study of diseases of infectious origin, that is, exogenous. What would Semmelweis think of our reckless use of antibacterial soaps when we had perfectly adequate soap and water (well, he did use lye to wash his hands so he may have been a bit overly fastidious)? Insidious infectious agents are clearly a dangerous external force and should be included in the definition of the exposome. Up until late in the 20th century, infectious disease and the environment were not comingled. It is more than the infectious agents being viewed as an environmental insult. Our environment and how we interact with it clearly influence the spread of disease vectors and our vulnerability to such vectors. Disease ecology addresses this important relationship and will likely contribute to our understanding of the exposome.

Moreover, our normal microbial flora or microbiome serves as a critical interface between our endogenous systems and our microbial surroundings. We know that environmental factors can alter our microbiome and that our microbiome can alter exogenous chemicals that enter our bodies.

From an investigatory perspective, genetics has a distinct advantage over the environment. The patterns of inheritance, gene replication, and regulation are wonders of nature and have been quite tractable (see Chapter 2, Genes, Genomes, and Genomics: Advances and Limitations), although it is ironic that the majority of the discussion of Darwin occurs among the geneticists when natural selection is primarily driven *by the environment*. The environment, or if the reader will allow, the exposome, is what is driving natural selection. Admittedly, the response of the organism and specifically its DNA is at the heart of the evolutionary process (i.e., the cumulative biological response), but Darwin is used in the classroom to lay the foundation for the field of genetics. What happened to the environment? Why is the *Origin of Species* (Darwin, 1859) not used to inspire students to study environmental health sciences? If one removes the external forces acting upon our genome, evolution would grind to a halt.

This brings us to the classical debate of nature versus nurture or genes versus environment. It is unfortunate that the word "versus" is even used when comparing the relative contributions of these two poles of the continuum. It should not a competition. If one accepts it as a competition, one must acknowledge that the gene side has the upper hand and this may be why it is a larger driver of biomedical education. However, it is not an "either/or" situation. Biology acts at the interface. There is a dynamic interaction between our genes and environment, with a complex level of interdigitation (more detail in Chapter 2, Genes, Genomes, and Genomics: Advances and Limitations). Admittedly, there is somewhat of a competition when one looks at resources at global, national and institutional levels. Thus if we are going to view this as a competition, the environmental side needs a stronger proponent. This is not to suggest that this is at the individual level, but rather at a conceptual level, and we must be better prepared to properly represent the environment. As noted in Chapter 2 Genes, Genomes, and Genomics: Advances and Limitations, a purely genetic approach to studying human disease is limited. The environment must be incorporated into our conceptual models of disease and health. The nature side of the discussion is very well represented. I would like to reaffirm that the exposome represents a quantifiable component of nurture. When viewed in this manner, the exposome assumes the role of conceptual advocate for the environment.

1.8 If it is so obvious, why this book?

If the association between external factors and human disease is so obvious, why do we need the exposome? Over the past century, modern medicine has produced some extraordinary advances. Surgical excision of tumors, repair of broken bones, development of antibiotics, noninvasive imaging techniques, robotic surgery, laser-based irradiation of tumors, immune-based therapies—for the most part these innovations have focused on fixing diseases after they develop. After all that is the model upon which our healthcare system is built, and research and development will deliver goods that fit into the model. While the dramatic external influences (car accidents, gunshots) are readily apparent, the subtleties of environmental influences have slowly slipped in recognition. Environmental health is not taught in most medical schools. Toxicology instruction is confined to the understanding of the adverse effects of the medicines prescribed by doctors, which is disconcertingly ironic. Doctors treat sick patients. They are certainly concerned with the obvious external forces that cause illness or injury (infection, trauma), and the less obvious external forces are underappreciated. Cumulative low-level exposures to chemicals in the workplace, home, or community are just not on the average physician's diagnostic radar. This is unfortunate, as we know long-term exposure to heavy metals, for example, can cause or exacerbate a variety of conditions. We know that many cancers are caused by exposure to toxicants in the workplace, but how often do you hear a physician inquire about one's environmental exposures? Smoking is an exception, but consider how acceptable smoking was within the medical community just in the past century. It has been estimated that if the practice of cigarette smoking (definitely part of the exposome) were eliminated throughout the world, it would have a more positive benefit to humankind than all of the advances of modern medicine combined (save vaccines). Control of the external forces represents a very powerful lever for improving health.

I am sympathetic to the physician, medical school faculty, and other allied health professions because it is exceedingly difficult to teach this material. There just is not a framework that allows appropriate integration into the current medical school and allied health sciences curricula. Enter the exposome. Perhaps I am overly optimistic, but I believe that the exposome provides the framework to educate medical and health professionals and inform the general public. I am not convinced that we need a new discipline of exposome medicine, but rather we need the medical establishment to consider how an inclusive concept such as the exposome can help patients integrate the complex information to which they are bombarded (more on this in Chapter 7, Pathways and Networks, and Chapter 8, Data Science and the Exposome). The exposome concept is not at all complicated as a concept. It is actually very simple. The cumulative impact of our exposures - we must consider them all.

As simple as the general *concept* of the exposome is, the measurement and interpretation of the exposome is quite a different issue. The level of technological and bioinformatic complexity needed for the exposome is daunting. As detailed in Chapter 7, Pathways and Networks, and Chapter 8, Data Science and the Exposome, the analytical and mathematical approaches needed to understand the exposome will exceed those used by the field of genetics. Even with the perplexing chasm that exists between the simple concept and the complex execution, there is attainable middle ground. Any movement from the simple concept to the complex elucidation is positive. Bringing more environmental health into human health provides an exciting opportunity to improve how we address healthcare. Moreover, an improved understanding of how the environment influences our health advances our understanding of biology.

1.9 The exposome community

One of the challenges of discussing a new research area is identifying the people who will drive the science. Where do those who have an interest in the exposome reside? Will they all be drawn from environmental health? Toxicology? Epidemiology? One can stratify these scientists based on their work setting and their discipline. Investigators interested in the environment generally work in academia, government, industry, or nonprofit agencies. Those in academia rely on the government for grant support and trainees for intellectual input and productivity. Governments support research and regulatory activities, which can have a major impact on public health. Industry scientists are often focused on minimizing the environmental impact of their particular sector or products. Those working for nonprofits are often focused on implementing practices that enhance the environment. For the most part, scientists interested in environmental factors in disease and health fall into one of three categories, which are covered in more detail in Chapter 4, The Environment: the Good, the Bad, and the Ugly. Those that study these relationships at the population level tend to be epidemiologists that specialize in the environment. Those that measure and evaluate the specific environmental exposures are considered exposure assessors or exposure scientists. Those interested in how the environmental contaminants impact specific biological molecules or pathways tend to fall in the domain of toxicology. These three distinct disciplines use different tools and approaches. While there is some overlap, the environmental epidemiologists are looking for interactions at a population level, the exposure scientists at the geographical, regional, or individual level, and the toxicologist at the molecular/cellular level (much work is done on a whole animal level, but rarely on an individual human).

In general, there is not a great deal of interaction among the major entities, but if progress is going to be made on the exposome, this must change. These three core disciplines, which exist among a wide variety of employers, will likely drive the field with the input of specialists in systems biology, bioinformatics, genetics (yes, genetics), chemistry, behavioral sciences and many other disciplines. Thus it is critical to get investigators outside the three core subdisciplines to view the exposome and its associated problems as an attractive scientific challenge. As noted in Chapter 7, Pathways and Networks, and Chapter 8, Data Science and the Exposome, the computational and bioinformatic approaches needed for the exposome are exceeding difficult but provide a challenging and potentially rewarding pursuit for an enterprising and mathematically inclined investigator. The exposome is an exciting opportunity for the exploding field of data science, and it is critical that those interested in exposome research engage with the data science community. It is incumbent upon those in the core disciplines to entice those with the necessary skills to collaborate on these projects or progress will be exceedingly slow.

As I write this, it has been 15 years since Wild coined the term. There has been an increasing amount of progress over that time. Much of this progress will be described in later chapters. The exposome is a growth industry. When I was writing the first version of this book in 2013, I recall writing with a sense of hope that the exposome would increase in

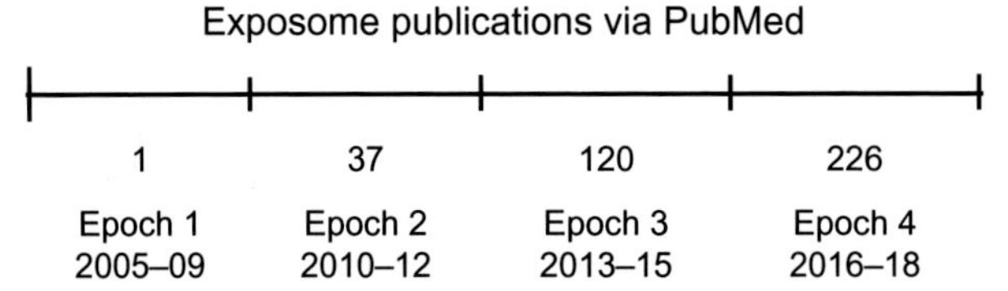

Figure 1.6 ***Exposome publications over time.*** A quick search of PubMed reveals the rapid growth in the field. Notably, the original Wild publication was the sole publication for many years. In 2020 the rate is approaching one paper per day.

recognition. It is safe to say that this has occurred. The efforts of research groups in the European Union and many other institutions are evidence of this growth. A simple analysis of PubMed reveals the upturn in publications that cite the word exposome (Fig. 1.6).

While it is still a relatively small number, the rate has increased from less than one publication per year, to one per year, to one per month, to one per week, to the current rate of nearly one per day. That is an encouraging acceleration.

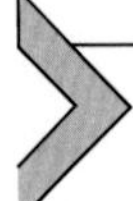

1.10 Tomayto, tomahto; apoptosis, a-POP-tosis; exposome, Xposome

As a graduate student, I bristled over the pronunciation of apoptosis as "a POP tosis." I was aware of the term ptosis for droopy eyelids with a silent p. Ptosis means to fall, apo meaning away-given apoptosis. So when heard that "pop," I cringed. I included a phonetic spelling of exposome in the first edition of this book and often lead discussions with the phrase "derived from the term exposure" to get people to emphasize the POZE, as in ex-POZE-ohm. My initial response to the Xposome pronunciation was similar to "a POP tosis" over 20 years ago. I recall the first meeting with Wild after I had written the book and was relieved when I heard his pronunciation (ex-POZE-ohm) with the extra bit of erudition from the British accent. I had heard him speak years before and thought that was how he pronounced it but had started to question my auditory recollections. Yet, I hear many other investigators in Europe and Asia use the Xposome pronunciation. Later, I came the realization that I was being overly pedantic, and different languages emphasize different syllables. That said, now I am comforted whenever the word is uttered, regardless of the syllabic emphasis, and I do not think we should quibble about this. The exposome is about exposures and as long as people are discussing it or writing about they can pronounce it however they would like to pronounce it.

1.11 Teaching/learning the exposome

As noted above with regard to physician education, the exposome can also be viewed as a vehicle for teaching the importance of the environment to our health. A single undergraduate course could help students place the myriad of exposures into the proper context and introduce them to historical and biological concepts and approaches. It forces the individual to consider all of the forces that are impacting their health, not just the obvious issues that were made apparent by trips to the pediatrician or the misinformation available on the internet. While genetics and socioeconomic backgrounds are nearly impossible to alter, as a young adult matures the activities, careers, behaviors, diet, and habits are under a considerable degree of control. Framing this within the context of the exposome provides an excellent foundation. I have received correspondence from faculty and students who have developed or participated in undergraduate or graduate courses that have introduced the exposome, and the feedback has been very positive. For the undergraduate, it provides an overarching concept for studying the environmental influences on health. This is especially important with the current level of disinformation swirling about social media platforms. For graduate students in toxicology, environmental health sciences, or epidemiology, the exposome helps place their research into the bigger picture of health science. It stresses the need to communicate with scientists outside their immediate discipline and the ever-increasing importance of big science within the scientific enterprise. Encouraging this type of collaborative research should occur early in the formative years of the science trainee. The exposure scientist must gain an appreciation for the biological mechanisms of disease and the complexity of population-based research. The toxicologist must understand how basic research is applied to the human condition. The environmental epidemiologist must assure that the associations are grounded in biological pathways. All must continue to explore disciplines that may lend new tools and approaches.

For the medical student, the exposome could provide a comprehensive model for consideration of environmental impacts on disease and health. When diagnosing a chronic condition, the physician must understand how environmental factors impact the patient and be able to explain the multitude of forces to the patient. The exposome may be able to provide this framework to enhance the doctor–patient relationship. For the established environmental health scientist, the exposome represents the future—the inevitable need to deal with the massive data sets that result from ever-advancing technology. While this flies in the face of the reductionistic approaches that are taught in graduate schools, it is imperative that we counterbalance our reductionism with thoughtful construction of complementary theories and hypotheses. Most scientists are reductionists by nature and produce "the bricks" (Forscher, 1963), but somebody has to assemble them. The exposome mandates that the process of building the structure is equal to that of generating the bricks. For the general public and community, the exposome provides a practical mental model for one's health as noted for the undergraduate student--an emphasis on the whole rather than the parts.

Regardless of one's place in the educational process (undergraduate, graduate student, teacher, professor, scientists, or casual learner), the exposome has value as a concept. The following chapters delve deeper into the practical issues that must be addressed to make the exposome valuable as a scientific construct. I encourage you to read the following chapters and suggested readings and supplement these activities by searching the current scientific and lay publications for information that can help inform our study of the exposome.

1.12 Obstacles and opportunities

I provide a section in each chapter on the potential obstacles and roadblocks that may be faced for that particular topic, as well as the opportunities and future directions. One could view the rest of the book as the obstacles and opportunities one may face from the comments in this opening chapter. As a new field is being established, its identity will face times of crisis. How are we defining the field? What do we want the field to be? The challenge for the exposome field is to define itself in a compelling and clear way and to then drive the science forward. The field must prevent offshoots and sects that dilute the message. We must work together to create a vision that inspires participants and instill confidence in those outside of the field. This is an opportunity that rarely occurs within the life of a scientist. I encourage the reader to join me and the rest of those interested in the exposome to seize this rare opportunity to define a field that is undoubtedly manifestly important. For now, let us not worry about the impossible.

1.13 Discussion questions

When was the first time you had heard to term genome? Exposome? Which do you think is more important to your own health? Why?

When you visit the doctor are you asked about family history or genetic influences of disease? Are you asked questions about your environment, diet, and habits? Do you think these questions capture the complexity of your exposome?

References

Buck Louis, G. M., Smarr, M. M., & Patel, C. J. (2017). The exposome research paradigm: An opportunity to understand the environmental basis for human health and disease. *Current Environmental Health Reports*, *4*(1), 89–98. Available from https://doi.org/10.1007/s40572-017-0126-3.

Darwin, C. (1859). *On the origin of species by means of natural selection, or preservation of favoured races in the struggle for life*. London: John Murray.

Forscher, B. K. (1963). Chaos in the brickyard. *Science*, *142*(3590), 339. Available from https://doi.org/10.1126/science.142.3590.339.

Kuhn, T. S. (1962). *The structure of scientific revolutions*. Chicago, IL: University of Chicago Press.

Kuhn, T. S., & Hacking, I. (2012). *The structure of scientific revolutions* (4th ed.). Chicago, IL; London: The University of Chicago Press.

Miller, G. W., & Jones, D. P. (2014). The nature of nurture: Refining the definition of the exposome. *Toxicological Sciences*, *137*(1), 1–2. Available from https://doi.org/10.1093/toxsci/kft251.

Patel, C. J., Bhattacharya, J., & Butte, A. J. (2010). An environment-wide association study (EWAS) on type 2 diabetes mellitus. *PLoS One*, *5*(5), e10746. Available from https://doi.org/10.1371/journal.pone.0010746.

Rattray, N. J. W., Deziel, N. C., Wallach, J. D., Khan, S. A., Vasiliou, V., Ioannidis, J. P. A., & Johnson, C. H. (2018). Beyond genomics: Understanding exposotypes through metabolomics. *Human Genomics*, *12*(1), 4. Available from https://doi.org/10.1186/s40246-018-0134-x.

Shakespeare, W., Orgel, S., & Braunmuller, A. R. (2002). *The complete works*. New York: Penguin.

Taleb, N. N. (2007). *The black swan: The impact of the highly improbable* (1st ed.). New York: Random House.
Taleb, N. N. (2012). *Antifragile: Things that gain from disorder* (1st ed.). New York: Random House.
Vrijheid, M., Slama, R., Robinson, O., Chatzi, L., Coen, M., van den Hazel, P., & Nieuwenhuijsen, M. J. (2014). The human early-life exposome (HELIX): Project rationale and design. *Environmental Health Perspectives, 122*(6), 535–544. Available from https://doi.org/10.1289/ehp.1307204.
Wild, C. P. (2005). Complementing the genome with an "exposome": The outstanding challenge of environmental exposure measurement in molecular epidemiology. *Cancer Epidemiology, Biomarkers and Prevention, 14*(8), 1847–1850. Available from https://doi.org/10.1158/1055-9965.EPI-05-0456.
Wild, C. P. (2012). The exposome: From concept to utility. *International Journal of Epidemiology, 41*(1), 24–32. Available from https://doi.org/10.1093/ije/dyr236.
Wild, C. P., Scalbert, A., & Herceg, Z. (2013). Measuring the exposome: A powerful basis for evaluating environmental exposures and cancer risk. *Environmental and Molecular Mutagenesis, 54*(7), 480–499. Available from https://doi.org/10.1002/em.21777.

Further reading

Rappaport, S. M., & Smith, M. T. (2010). Epidemiology. Environment and disease risks. *Science, 330*(6003), 460–461. Available from https://doi.org/10.1126/science.1192603.
The National Academies. The exposome: A powerful approach for evaluating environmental exposures and their influences on human disease. Washington, DC: The National Academies; 2010. <http://nas-sites.org/emergingscience/files/2011/05/03-exposome-newsletter-508.pdf>

CHAPTER TWO

Genes, genomes, and genomics: advances and limitations

It is essentially immoral not to get the human genome sequence done as fast as possible.

James Watson

2.1 DNA, no longer a secret

The discovery of the secret of life as announced by James Watson and Francis Crick at the Eagle Pub in Cambridge ranks high in the annals of scientific history. I had the pleasure of visiting the Eagle a few years ago, and it was one of the most memorable scientific field trips I have taken, and for good reason. The elucidation of the structure of deoxyribose nucleic acid (DNA) stands out as one of the greatest scientific discoveries of all time. As stated in the opening lines of the 1953 *Nature* article, "we wish to suggest a structure for the salt of deoxyribose nucleic acid (DNA). This structure has novel features which are of considerable biological importance" (Watson & Crick, 1953b). This surprisingly reserved statement from the pair, not known for being subtle, foreshadowed a revolution in biology. It is difficult for scientists in the 21st century to fathom life without knowing the structure of DNA. It is taught in grade schools, but just over 60 years ago the term "double helix" was not in the scientific vernacular. Although the concept of the gene existed and there was evidence that DNA was likely the genetic material, there was no understanding of how it transferred the information. The penultimate line penned by Crick was similarly prophetic; "It has not escaped our notice that the specific pairing we have postulated immediately suggests a possible copying mechanism for the genetic material." In follow-up manuscripts, Watson and Crick (1953a) describe the pairing mechanism that allowed for the DNA to serve as a template for gene copying. These findings represented a paradigmatic shift in the truest Kuhnian sense. The

The Exposome.
DOI: https://doi.org/10.1016/B978-0-12-814079-6.00002-X

auctioning of Francis Crick's Nobel Prize ($2 million USD) and letter to his son describing their discovery ($6 million USD) bears witness to how extraordinary this discovery is viewed in modern scientific history. Of course, many other scientists contributed to the knowledge that permitted the discovery—Wilkins, Franklin, Bragg, Pauling, Kendrew, and Perutz to name a few. Of course, the two other papers from that *Nature* trifecta that included the key X-ray crystallography data from Rosalind Franklin were critical to the advancement (Franklin & Gosling, 1953; Wilkins, Stokes, & Wilson, 1953). It was a stunning discovery based on a combination of thinking and tinkering that would forever change biology. The impact of this discovery and the implications of it would transform biomedical research.

2.2 The Human Genome Project: DNA at scale

Move forward 50 years to the completion of the first draft of the Human Genome Project in 2003. This was not a handful of scientists working with X-rays and tinkering with models. The Human Genome Project was big science– billions of dollars, hundreds of scientists, and three billion base pairs. Although the 1986 Santa Fe conference is often considered the kickoff to the human genome sequencing project, one might say it started back in Cambridge. The Human Genome Project required some major technological advances (see Fig. 2.1). It is remarkable to think that when the Human Genome Project was first being discussed, the existing technology was rather slow. As such, the project would have taken over 100 years to complete had there been no improvement in sequencing technology. Indeed, without the development of sequencing by Gilbert and Sanger (for which Sanger won his second Nobel Prize) and its further automation by Leroy Hood, there would be no initiative. This technological advance spurred scientific water cooler discussions of sequencing the entire human genome. The often underappreciated efforts by the U.S. Department of Energy, with its interest in studying the mutagenic effects of radiation, played an important role in generating key infrastructure and knowledge. Technology was key. Another example of a critical technological advance was the discovery of the polymerase chain reaction (PCR). Most scientists take the routine technique of PCR for granted. Scientists had been working with DNA for decades, but the

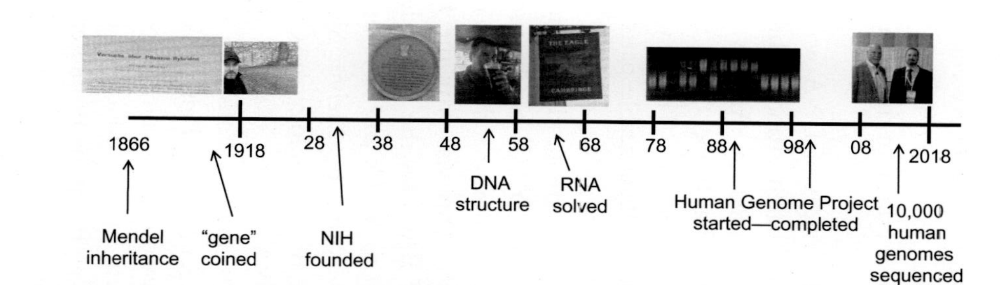

Figure 2.1 A century of genetic progress. For over 100 years there has been steady progress punctuated by a series of major discoveries with a corresponding level of accolades. Over the past few years, DNA sequencing service is available to the consumer via companies such as 23andMe and Ancestory.com. Many of the locations of these accomplishments are excellent scientific field trip destinations: author at Mendl Museum, Eagle Pub, and with Craig Venter (from left to right). *DNA*, Deoxyribose nucleic acid.

ability to produce large quantities of specific sequences was technically challenging. Enter *Thermophilus aquaticus* and a quirky, some say bizarre, surfer/scientist from California. Kary Mullis was a scientist at Cetus Corporation, where a group of investigators was working on methods of DNA copying. But it was Mullis who, not unlike Watson and Crick, assembled the *ex post facto* apparent pieces to solve the riddle (and be awarded a Nobel Prize). Mullis attributes some of his creative thinking to his prior (heavy) use of LSD, which may also have contributed to some of his subsequent pseudoscientific ideas. However, it is also highly likely that input from other Cetus scientists was critical in the formation of his breakthrough discovery.

Being able to manipulate large pieces of DNA was enhanced by the development of bacterial and yeast-based artificial chromosomes (BACs and YACs). YACs were first described by Andrew Murray and Jack Szostak. Szostak went on to win the Nobel Prize in Physiology and Medicine for the co-discovery of telomeres. These systems allowed the genome to be fragmented into more manageable parts and be grown in cell-based systems. YACs are composed of an artificial chromosome that contains a centromere, telomeres, and a replication origin element, which allows replication in yeast cells. While YACs can be used to clone fragments from 100 to 3000 kb, they unfortunately have a tendency to be unstable and can introduce errors. This was why the Human Genome Project transitioned to the use of BACs for the remainder of the project. BACs have a similar composition to YACs and have become a mainstay in the study of genetics. First reported by Melvin Simon and Hiroaki Shizuya at the California Institute of Technology, BACs are efficient vector systems for chromosomal DNA libraries (Kim et al., 1996). Besides their utility in sequencing projects, many scientists are also aware of their utility in the production of transgenic mice. Being able to include much larger sections of DNA than previously possible allows large sections of genomic DNA to be transferred to the donor embryonic stem cell, including the ability to employ endogenous promoters to drive expression. This is especially helpful when modeling complex genetic disorders that involve regulatory units that span large distances on the chromosome.

Armed with the aforementioned significant technological developments and support from international heavyweights, such as the Wellcome Trust in the United Kingdom and the U.S. National Institutes of Health and Department of Energy, the field moved forward with this Herculean project. Major centers involved in the project include the

Whitehead Institute, the Wellcome Sanger Institute, Washington University in St. Louis, and the Baylor College of Medicine. The NIH side of the effort was headed up by none other than James Watson. Watson provided a high level of enthusiasm and gravitas to the project. Over the course of the next several years, swift progress was made. The government-funded initiative was proceeding slightly ahead of schedule. Then enter one Craig Venter from The Institute for Genome Research (TIGR) and his newly formed company Celera Genomics (celera means swiftness in Latin). See James Shreeve's book *The Genome War* (Shreeve, 2004) for an excellent overview of Venter's contributions. The brash scientist, who had previously worked on the Human Genome Project while employed at NIH, proposed to use a shotgun approach to complete the sequence of DNA in a mere three years (four years faster than what the Human Genome Project had planned). Who says science isn't fun? His company, Celera Genomics, had raised a substantial amount of private funding and proposed to do the job better and faster. As a budding scientist, I recall the race for the sequence of DNA and was entertained by the sheer audacity of Venter. While there was concern that Venter's approach was scientifically more coarse and perhaps professionally more crass, it was clear that a sense of urgency had developed among the Human Genome Project teams. This is a wonderful illustration of the value and travails of competition in science. Eventually, Venter and his company begrudgingly partnered with the Human Genome Project team, then led by current NIH Director Francis Collins, to complete the project together. However, the acceleration of the project would not have occurred without the tacit and overt sense of competition from Venter's organization. It was great scientific theater. I had the pleasure of interviewing Venter when he spoke at the Society of Toxicology meeting in 2015. We discussed some of the work on the exposome, and he seemed genuinely intrigued with how environmental information could be extracted from untargeted metabolomic/mass spectrometry approaches. If only we could get him to marshal resources for exposome research.

One of the issues surrounding the efforts of Celera Genomics was the intent to patent genes and genes sequences. This was something Watson opposed with a passion. Indeed, Watson stated that patenting DNA was lunacy. In 2013 the U.S. Supreme Court agreed with Watson by nullifying Myriad Genetics' patent on the gene associated with breast and ovarian cancer, BRCA1, but left the door open to the patenting of DNA sequences not found in nature. Ultimately, this disagreement led to

Watson's resignation in 1992. Francis Collins soon took over as director of the Human Genome Project. After a series of bigger-than-life DNA personalities, it has been comforting to see a seemingly even-keeled and reserved Collins get the Human Genome Project to the finish line and become the leader of the National Institutes of Health. But I will note that the ongoing genome-centric mindset of NIH must be attributed, in part, to Collins' background and genetic predilections. He has signaled some interest in the exposome, but to date the resources and energy have been tepid. I believe that some of the reservations have been due to the lack of coordination and focus among the exposome community. Hopefully, in the near future the efforts outlined in later chapters will garner more resources from NIH. Collins has compiled a list of important lessons learned from the Human Genome Project. In a paper titled *The Human Genome Project: Lessons Learned from Large-Scale Biology*, Collins made it clear that big science is different from the traditional way of doing science. Indeed, this paper, which I revisit in the closing chapter, is a gift to the exposome initiative. Many of these lessons are directly applicable to steps that would need to be followed to pursue a human exposome project. Let us hope that NIH leadership expands its support of exposome research. The past and current activities are discussed in Chapter 10, The Exposome in the Future. Linda Birnbaum, the director of the National Institute of Environmental Health Science from 2009 to 2019, was an enthusiastic supporter and helped get Collins to discuss the concept of the exposome with NIH leadership. I heard Gary Gibbons, director of the National Heart, Lung, and Blood Institute, speak at a conference that had nothing to do with the environment, and he said the word exposome four times in his talk. Christopher Austin, director of the National Center for Advancing Translational Sciences (NCATS), has been a long-time supporter of the concept and participated in the critical 2010 National Academy of Science workshop on the exposome. As NCATS oversees the Clinical and Translational Science Awards, there is an excellent opportunity to get more exposome-related research embedded into the clinical research enterprise. The fact that several NIH leaders are warming to the concept bodes well for the future of the exposome. NIH has made substantial advances in our understanding of our genome. The exposome can help extract more information out of our genome by acting as a foil.

Oxford English Dictionary: Foil (noun):

> *A person or thing that contrasts with and so emphasizes and enhances the qualities of another.*

2.3 Heritability

Given that our phenotype is a result of our genome and exposome, it is important to understand the genetic contributions. In fact, I believe that elucidation of the exposome will heavily rely on the genome and genetics to frame our boundaries, guide our large-scale biology, and provide inspiration for such a grand endeavor. One of the most important concepts for this is heritability. This is a topic that is taught in grade school and is intuitively obvious to the average citizen. We know that children often resemble their biological parents. Hair color, eye color, body shape, and mannerisms are all features that run in families. But it is not as simple as that. Families also tend to share environments, and the way that parents nurture their children has a major effect on many traits (more on this in Chapter 3, Nurturing Science). Mannerisms such as facial expressions have a strong environmental component, whereas traits such as eye color are more fixed. Thus, we must first dissect the component of a trait or disease that is truly heritable. For this, we must examine how this is determined. By defining the genetic influences on biology, heritability will set the boundaries for where the exposome has an opportunity to influence outcomes.

First, we must consider the nature of the outcome we are measuring. In some instances, the outcome is binary- that is, either the person has or does not have a particular feature (for example, having a window's peak hairline or unattached earlobes). Other traits have discrete distributions such as eye color: blue, gray, brown, green, and hazel. Then many others exist on a range— for example, height and weight are considered to be quantitative.

Peas and monks—not your typical field trip

As part of my ongoing scientific sojourns, I was able to visit the Muni Mendel Museum in Brno, Czech Republic [officially, the Mendel Museum of Masaryk University (MUNI)], where Gregor Mendel conducted his pioneering work on the inheritance of traits. His work focused on pea plants and was conducted on the grounds of the St. Thomas Monastery where he worked (at that time part of Moldavia). Several years ago, MUNI adopted the museum and currently serves as the steward of the exhibition. Visitors are not limited to sterile museum exhibits, as they can walk across this hallowed ground, literally and

(*Continued*)

(Continued)

figuratively—as it is still the site of the monastery. Mendel's meticulous descriptions of the crossing patterns and offspring can be viewed in his original texts on display at the museum. It was unfortunate that his work went unrecognized for decades. Darwin was very much embedded in the scientific circles of the day and his ideas were rapidly disseminated. Darwin was not privy to Mendel's work that essentially provided the mechanistic basis for the natural selection he described. Had he been aware, Darwin's work could have had an even greater impact. I was pleased that the opportunity to visit Brno was not due to a conference on genetics or genomics, but rather an opportunity to speak about the exposome at MUNI within their RECETOX program, which is becoming a major player in the exposome space under the leadership of Jana Klánová.

In the case of human disease, there are four primary means of estimated heritability. The first is a method based on the correlation of disease states in pairs of relatives selected from a random sample of a population. The second is the commonly used twin method in which a disease is studied in populations of monozygotic (identical) and dizygotic (fraternal) twins. Generally, either type of twin shares the same prenatal and postnatal environment, that is, same womb, same household, and same diet. However, the monozygotic twins have identical DNA, whereas the DNA in the dizygotic twins will differ (although they still have the same parents and will share large portions of their DNA; see Fig. 2.2). The third approach is called Falconer's method, after the Scottish geneticist Douglas

	r	s
Monozygotic twins	1	1
First-degree full sibling	0.5	0.25
Parent–child	0.5	0
Second-degree child–grandparent	0.25	0
Third-degree cousins	0.125	0

Figure 2.2 How similar are we? Our genetic makeup comes from our parents, but that percentage of DNA that we have in common with our relatives can be dramatic. This figure shows the predicted amount of shared DNA among various biological relatives. r is equal to the coefficient of the relationship, with s being the probability of two relative pair having the identical genotype. *DNA*, Deoxyribose nucleic acid. *Adapted from Tenesa, A., & Haley, C.S. (2013). The heritability of human disease: estimation, uses and abuses.* Nature Reviews Genetics 14*(2), 139–149. doi:10.1038/nrg3377.*

Falconer. It was first introduced in his 1960 book on quantitative genetics (Falconer, 1960). It examines correlation in disease liability in relatives from the incidence of the disease in the general population and among relatives of individuals with the disease. The fourth and last method is a Bayesian-based approach that uses generalized mixed linear models.

Thus, the simple Punnet squares learned in biology and genetics courses, which are superb for explaining the concept well described by Mendel, merely scratch the surface of the complexity of heritability of human disease. A thorough understanding and appreciation of the causal association between genetic variation and human disease and traits is essential for exposome research. Steve Rappaport has noted that the exposome is essentially 1 − G (1 representing the phenotype, G being genetics). Master the limits of the genetics and the rest of the solution will be found in the exposome.

Exposome research is rightly focused on the nonheritable portion of disease causation, progression, and treatment, but it is more than just the nonheritable portion. Specifically, the environment-related variance portion of the analysis can be referred to as the *preventability*— that is, the portion of the variance that has thc potential to be altered (Tenesa & Haley, 2013). This is the crux of prevention, as it relates to public health and also has major implications for personalized medicine. We want to identify and act upon the preventable contributors to disease and other adverse health outcomes, instead of focusing on fixing diseases after they develop. I revisit this concept in Chapter 9, The Exposome in the Community, as part of the modifiable exposome.

To help illustrate the relative contributions of genetics and exposome on health I am going to introduce the reader to the "Example" family. I will use the members of The Examples throughout the book. The Example family is composed of a mother, father, and eight children (Fig. 2.3). Two of the children are monozygotic twins (boys, age 10, Edward and Eustas), two are dizyogotic twins (one girl and one boy, age 8, Emily and Elwood), two additional biological children (Elmer 6, Eloise 4), and then two girls (Ester 12 and Esmerelde 5), who were adopted. It is a large family and one with intentional genetic variety (Fig. 2.4).

So what can genetics tell us about the Example family? Six of the children should share 50% of their parents' DNA, but in different combinations. Two of the children are adopted and share DNA with no other family member. In their day-to-day life the genetic blueprints for each family member dictates many critical health factors. Genes can influence

Figure 2.3 Introduction to the example family. Parents Ernest and Eliza live with their eight children. The genetic background of their children are described in Fig 2.4.

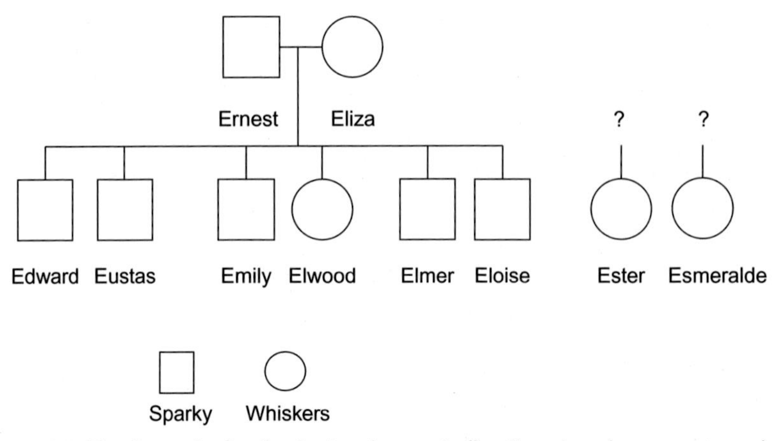

Figure 2.4 The Example family depicted genetically. Genetics plays a major role in one's vulnerability to disease, but even within a family, there can be significant variation among members.

the biological response to food, medicines, and even social stressors. Their height, hair color, and disease vulnerability are all influenced by their genetics. Yet, the Example family has quite a bit of genetic variability.

The biological children are inheriting approximately half of their genes from mother and father (in a somewhat random fashion). Some twins have identical DNA (identical twins or monozygotic) while the others (fraternal or dizygotic) are different. Some have shared the same in utero environment; others have not. At this time, they all reside in the same household and go to many of the same places. I address their shared and unique environments in Chapter 3, Nurturing Science, and Chapter 4, The Environment: the Good, the Bad, and the Ugly, but let us examine their genetics a bit more closely, as well as the general heritability of disease (Fig. 2.5).

For a holiday gift, the Example family decides to submit their salivary-based DNA samples to a company for genome analysis. After a few weeks they are provided with 10 reports. The analysis confirms that six of the Example children are indeed Ernest's and Eliza's biological children. Ester and Esmeralde are shown to be genetically unrelated to the other six family members. The twins, Edward and Eustas, are indeed identical, whereas the twins Emily and Elwood are confirmed to be fraternal. Elmer and Eloise are genetically related to Ernest and Eliza, but there is a great deal of mosaicism and there is little overlap. Ernest's family has a history of macular degeneration and Eliza's family has a history of heart disease. Some of the expected risk alleles are present for both of these disorders.

What is the likelihood that a reported finding changes the behavior or routines of any of the family members? If you find that you have a slightly increased risk of macular degeneration, would you be more likely to wear sunglasses? Would a teenager be able to fathom that her current activities are linked to a disease that will not occur until her 70s? Was it necessary

Heritability of major diseases	
Depression	0.576
Type II diabetes	0.561
Non melanoma skin cancer	0.520
General hypertension	0.462
Asthma	0.457
Allergic rhinitis	0.445
Osteoarthritis	0.256
Eye infection	0.200

Figure 2.5 Heritability of several human diseases. Heritability is scored on a scale of 0–1. A score of 1 means that all of the variance in the disease can be explained by genetics. A score of 0 means that none of the variance can be explained by genetics. The majority of human diseases cluster in the 0.3–0.7 range, indicating the importance of our genome and exposome.

to have the genetic testing given that the family had long known about the macular degeneration and cardiovascular disease in the family tree? Thus, we can generate a great deal of information about the genetics of the Example family, but how much of that information is translated into actionable decision-making? If the parents carried risk alleles for Tay–Sachs disease, phenylketonuria (PKU), or cystic fibrosis, the genetic information would be highly relevant for immediate medical intervention, but for chronic disease, there is less actionable intelligence. It is possible that the results make the concern more personal for the individual, but it is not clear if this will lead to long-term changes in behavior.

This is not to say that genetics lacks value in understanding human disease and health. We have learned a great deal about biology from genetic research. One of the more powerful examples comes from genome-wide association studies (GWAS), hailed as a novel way to identify genetic associations with various disorders. By analyzing the genetic variation in populations (by DNA microarray technology), it is possible to relate gene mutation to a particular health or disease outcome. Data are displayed in the familiar Manhattan plot design where the landscape shows associations by taller peaks at given chromosomal locations. In seconds the human mind can scan the association with the aid of the colorplot. There have been numerous successes with novel associations and pathways identified, but the relatively low-resolution of single nucleotide polymorphisms (SNPs) has limited the utility of the GWAS. Many studies have been criticized for study design and lack of appropriate clinical characterization of the patients, but these concerns hold true for any type of research conducted on a human population. A 2012 opinion piece in the *Journal of the American Medical Association* lays out the promises and limitations of GWAS, but foreshadows its potential demise (Klein et al., 2012). As another example of the swiftness of science, the authors suggest that exome or deep sequencing, where the complete sequence of suspected chromosomal region is obtained, will soon supplant the GWAS (although some of the same statistical approaches will be used). I predicted that full genome sequencing would be available and affordable in the first version of this book and in fact it is now readily available. Availability of every person's genome could be viewed as the true completion of the Human Genome Project. Yet, even armed with the complete genome, major impediments to understanding disease persist. Three billion base pairs—that is a big number. The computational power required to conduct such analyses continues to grow and strategies to tame these types of data sets are discussed in Chapter 7, Pathways and Networks.

One of the most important findings of the large-scale GWAS studies is that many genetic variations contribute to a particular disease outcome. For example, there may be 10 genes that can increase or decrease the risk of a disease. Some have proposed the use of polygenic risk scores to integrate the relative risk of each gene into a single score. It sounds promising, but it is not clear if the underlying statistical approaches are valid (Janssens & Joyner, 2019). Moreover, since there is underrepresentation of minority populations in GWAS databases, there is concern that polygenic risk scores could exacerbate health disparities (Roberts, Khoury, & Mensah, 2019). The development of these approaches is important for exposome research, as the exposome may ultimately be a poly-exposomic score. As noted above, the exposome field will benefit from a close examination of the advances and missteps in the genetics community.

Another key genetic development was the International Haplotype Mapping (HAPMAP) project. In this initiative, started in 2002, the research team set out to assess the variability in the human genome. Since the Human Genome Project involved the sequencing of a small pool of human DNA, the initial sequencing could not address the diversity among the human population let alone how those variations contributed to human disease. Therefore the HAPMAP project aimed to assess how much the human genome varied across people throughout the world by sampling a larger and highly diverse population for analysis (from an original ancestor approximately100,000 years ago). SNPs represent areas of the DNA that are variable. An SNP chip may have 1 million different SNPs, but this represents a mere fraction of the genome. By obtaining data on the relative incidence of SNPs among different human populations, it has been possible to identify points of divergence and various disease associations. This even includes the determination of one's percentage of Neanderthal DNA. When I wrote the first version of this book 23andMe claimed I was in the 94th percentile, but as I prepared the second version, my percentage had dropped to 64%. Have I evolved in six years or have the genetic techniques evolved? Although I prefer the former, I am pretty sure it is the latter, but this shows the vagaries of science. Technologies improve and results can change accordingly. Some may fault the reporting of updated results as a lack of reproducibility, but in fact it is part of the scientific method. Even as I extoll the virtues of the power of genomics, the field still struggles mightily with its own limitations, which serves as a reminder that the data we have that points to the level of "fixedness" in the genome is subject to change (not the actual fixed nature, but the data reporting it).

One of the concerns with GWAS- and SNP-based studies is that SNPs shown to be associated with a certain condition may be misleading. Linkage disequilibrium is where two SNPs have been inherited together because of their chromosomal proximity. Thus SNP 123456 may be responsible for an increased incidence of disease X, but not be on the SNP chip (because only a small portion of the genome is represented). It is possible that SNP 123000 and SNP 124000 have been inherited together with 123456 over time and, thus, show up positive for the association with disease X, but only because they are located near the culprit SNP 123456. Not having SNP 123456 on the chip and only having information from the flanking SNPs would yield a misleading low-intensity effect with a low odds ratio (OR). But if the experiment were repeated with a chip that contained 123456, the OR would rise significantly. This is an argument for deep (or complete) sequencing, where the entire portion of that particular gene is sequenced. The main point is that SNP-based approaches do not provide a high enough degree of resolution to pin down some specific associations.

As of this writing the companies 23andMe and Ancestory.com will measure over one million SNPs for less than $200 USD. Over 20 million people have paid for this service. Risk alleles for disease conditions can be determined by submitting a sample of saliva. 23andMe uses the same type of SNP chip that has been used in GWAS studies. For example, if a study shows that people with a particular allele have a 20 percent increased risk of disease, 23andME would assign a 20 percent increased risk *based on that particular allele.* 23andMe is unable to assess other factors that impact that disease. Statistically, 23andMe is correct in assigning this 20 percent increased risk, but this is accurate only at a population level. A person with this inherited 20 percent increased risk may have a handful of positive risk factors that actually make that individual 30 percent less likely to get the disease. The provided values are only measuring genetic risk, and the company repeatedly reminds the consumer of this, but the participant does not have access to an equivalent value of their environment or risk factor impact. Individuals can be provided with genetic cardiovascular risk factors, but if the consumer does not have access to information on what actions can influence these risks, the impact on health is minimal. Ideally, the consumer would be provided with an analysis of the modifiable risk factors and how they can counterbalance the genetic risks, but at this time this type of analysis is not available.

However, 23andMe does allow one to download their raw data. That data can then be uploaded to websites, such as Promethease, which unveils all of one's genetic variants—no filters. Your genomic data should be available to you, but how many consumers have the genetic know-how to interpret this type of information? There is a reason that genetic counseling has developed as a field and that trained genetic counselors deliver results from genetic tests in the clinic. I appreciate FDA's restriction on the diagnostic nature of genomic studies, but consumers should have full access to their genetic information, and that means it is being delivered in a form that they can understand. Although my view on this stems partially from a consumer advocacy standpoint, my exposome-centric persona believes that when consumers have this comprehensive genomic information, they will come to the realization that their future lies in the exposome. That it is not genetic manipulation that will improve their lives, decrease disease incidence, and increase longevity, but rather the exposome that holds the key. Would it not be great to have a corresponding level of information on our environmental exposures? At the same time, we must be wary of how this information can be misused or misinterpreted.

2.4 The gene versus environment continuum

It is relatively easy to evaluate the causes of disease at the two ends of the continuum. Autosomal dominant or autosomal recessive diseases give rise to conditions that have been followed through family pedigrees for centuries. The Habsburg Dynasty of European rule is an interesting illustration. The majority of their lineage derived from genetic inbreeding and led to the infamous Habsburg family mandibular prognathism that was manifested in their unique jawline and lip features. Ultimately, the male offspring were infertile (an independent genetic problem unrelated to the mandibular peculiarity) and the royal and genetic lineage ended. This is an extreme trait perpetuated through selective breeding and one that the environment played little role. On the other end of the spectrum, conditions due to completely external forces, such as head trauma, asphyxiation, or even shark attacks appear to have no genetic component. The role of one's genetic background in response to a car accident is minimal, although the genetically endowed reaction time, visual acuity, and physical and mental dexterity may play a role in the cause of the incident, and clotting profiles may impact the response and survival, but for the

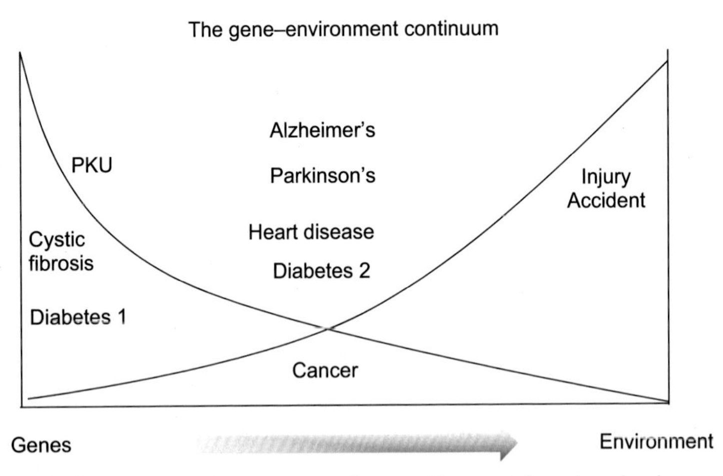

Figure 2.6 The gene–environment continuum. The past few decades have generated superb data on the genetic causation of disease. Numerous diseases result exclusively from genetic abnormalities. These are depicted on the left side of the graph. There are also outcomes that result exclusively from external or environmental sources, shown on the right side of the graph. The vast majority of diseases though, reside at the interface. They have genetic components that predispose the individual to the disease but are altered or accelerated by exposome influences. In order to address the majority of diseases at the interface, we must have more comprehensive environmental data (i.e., the exposome).

most part the incident is environmental. Figure 2.6 illustrates the relative contributions of genes and the environment to several diseases.

Back on the genetic side of the continuum, there are some instances where the environment does play a role. The age of onset of disease can vary among these genetic disorders, perhaps through environmental influences. An interesting example is PKU. PKU results from the mutation of the gene that metabolizes phenylalanine, one of the 20 amino acids that serve as building blocks for our proteins. Until 60 years ago, PKU was a leading cause of mental retardation. Once it was discovered that the disease was the result of altered amino acid metabolism, the medical field was able to develop a test and treatment. While it is not trivial to follow, the treatment is somewhat simple: severely limit the intake of phenylalanine in the patient's diet. Or to put it another way, *modify the patient's environment*. This insanely strict diet followed with precision entirely prevents the adverse effects on brain function. A mutation in a single gene causes a very serious disease that can be completely treated by altering the environment. While this stands out as one of the clearest examples of the ability of the environment to trump the genome, the formula may hold true for many other disorders.

The picture gets cloudier when we move away from the extremes. Complex disorders such as heart disease, diabetes, and Alzheimer's disease have multiple genetic and environmental contributors. There could be a dozen different genes that confer a particular level of risk in an individual and dozens of environmental factors that interact with those genetic factors. As noted above, the field of genetics is making great progress on identifying the multiple genetic factors that may contribute to these diseases, but without a similarly detailed analysis of the environmental contributors the puzzle will remain unsolved. The study of complex human diseases must examine genetic and environmental influences and do so in a systematic fashion.

2.4.1 A dangerous metaphor?

The balance between genes and environment has often been summarized by a common phrase bandied about in environmental health sciences. I have personally heard it from two directors of the NIEHS, as well as Francis Collins, director of the U.S. NIH. The first reference to it occurred in the 1920s by Elliot Joslin at Harvard Medical School (Fig. 2.7).

I have never been fond of this phrase in that is has a very fatalistic tone. It is as if the only thing the environment can do is pull the trigger (and uses an unfortunate gun violence phraseology). It suggests that all environmental influences lead to catastrophic damage. This is not the case. Our environment can have beneficial effects on our health and can even contribute to the repair after damage. Ironically, Joslin was a key proponent of environmental modification as medicine, and the Joslin Diabetes Center at Harvard University has been a leader in the use of environmental modification to treat the disease. More important, however, is that careful regulation and manipulation of our environment can tie a knot in the barrel, rendering the genetically endowed predisposition

"Genetics loads the gun,
environment pulls the trigger"

Figure 2.7 The gun analogy. This analogy has been used for many years to describe the interaction between genetic and environmental factors. Perhaps because of the difficulty in modifying our genetics, the environment has been viewed from the same static perspective.

"Genetics loads the gun and the environment may or may not increase the likelihood of the trigger being pulled, but the environment could also influence whether or not the bullet remains in the chamber or in the event that the trigger is actually pulled the environment could still influence the path of the projectile and even be involved in the response to the impact."

Figure 2.8 An expanded and intentionally absurd version of the gun analogy. The environment is not a unidirectional influence. Our external environment, or exposome, can be altered in many ways. Many genetic diseases are treated by manipulating the patient's environment. There are multiple reasons to stop using the gun analogy.

essentially moot. Our destiny is not a loaded gun. Adoption of practices that minimize our exposure to adverse factors and maximize our exposure to positive factors can have major impacts on our health. The more appropriate phrase is shown in Figure 2.8. Alternatively, we could just stop using the original phrase and focus on the modification of our environment in a way that makes us healthier.

2.5 Large-scale genetics

Let us move forward, say 20 years, when it is possible and affordable to conduct a study on a population of 10,000 with complete genome sequencing. We know all 3 billion base pairs for each of the 10,000 participants. Based on our knowledge of genetics, we are able to predict (with indeterminate accuracy) which people are more likely to develop diabetes, cancer, or heart disease based on their genetic risk. But what then? Let us refer back to Darwin's natural selection. Natural selection is a response of the species (and the DNA) to external forces. It is the external forces that drive the variation. Thus in our study population of 1,000, we can make all sorts of predictions of who will get what. The genetic data are fairly precise. But we know that all of the predictions are statistically based and assume a normal distribution. Yes, on average a certain allele will increase the incidence of x within a certain population, but it is not *ipso facto* for each individual. Manipulation of one's environment can alter one's presumed DNA destiny. The predictions will fail at a fairly high rate. Why? Because of the individual variability exerted by the external factors that make up the exposome. External forces acting upon the individuals will alter their genetically predetermined path. The only certainty is that these deviations will occur. When we observe differences in disease incidence based on geography, it strongly supports the idea that environmental factors are the culprit (as long as the population is genetically diverse).

The Encyclopedia of DNA Elements or ENCODE project was launched in 2003 by the U.S. National Human Genome Research Institute (NHGRI), The Wellcome Trust (United Kingdom), Riken (Japan), and the Universities of Geneva and Lausanne (Switzerland). The goal was to determine all of the functional elements of the human genome, including the genes, transcripts, transcriptional regulatory elements, chromatin states, and methylation patterns. All of the data generated from the different institutions, including gene annotation, transcriptome analysis, chromatin analysis, transcription factor binding, and methylation, undergo a verification and metadata process that ensures data integrity and usability. A massive amount of data from this project is now freely available online (http://genome.ucsc.edu/ENCODE/) with multiple search features. Many of the suggestions made by Collins, Morgan, and Patrinos (2003) (from the Lessons learned paper) have been wisely applied to the ENCODE project- namely, international involvement, clear goals, and staging. I revisit these concepts in the last chapter, but it warrants repeating. There are excellent examples of large-scale science that have been conducted in a scientifically rigorous manner. Proponents of the exposome should closely examine these strategies.

2.6 Epigenetics: a clear gene–environment interface

The epigenome is also addressed in the next chapter, but an introduction to the topic of epigenetics is appropriate here. While it was initially thought that the primary sequence of the DNA was the driver of heredity, it was quickly discovered that there were additional levels of regulation above that of the primary sequence. These modifications of DNA generally fall under the domain of the field of epigenetics/epigenomics. Epigenetics refers to all of the heritable changes in gene expression that are not coded in the DNA sequence itself but result in an altered phenotype without changing the genotype. There are three primary mechanisms of epigenetic regulation currently known: DNA methylation, histone modification, and noncoding RNA-mediated silencing. Epigenetic regulation represents an intermediate process that imprints information from

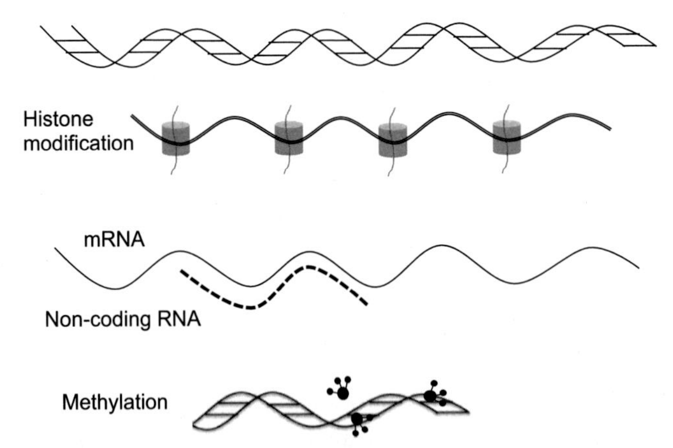

Figure 2.9 Mechanisms of epigenetic modification. Even with the same primary sequence genes can differ in their ability to be transcribed via epigenetic alterations. Histone modification alters the accessibility of the genes to the transcriptional machinery. Noncoding RNA can block regions of mRNA essentially blocking its translation (the basis of siRNA). Addition of methyl (or similar) groups to the nucleotides at CpG islands can effectively turn on or off a gene. Methylation is especially important in the context of the exposome in that chemical exposure can alter the level of methylation.

environmental experiences on the genome. Thus, the apparently fixed genome can be modified in a way that results in stable alterations in the phenotype (See Fig. 2.9).

When cytosine (C) is connected to a guanine (G) with a phosphodiester bond (p) in the DNA, it is called a CpG site or island, and it can be methylated by DNA methyltransferases. Methylation of these sites can result in silencing of the gene by repressing transcription. Aberrant CpG island methylation is reported to occur early in the process of tumorogenesis. Histones can undergo multiple covalent modifications that can regulate transcription including acetylation, methylation, and phosphorylation. While many of these processes are part of our endogenous regulation of DNA, most can be directly impacted by environmental exposures. Thus, epigenetic regulation can be viewed as the most direct interface of our genes and our environment. Geneticists may argue that the epigenome is part of the genome, and they are correct. But the modification of the genome through epigenetic mechanisms can clearly be considered as a result of an external force. The Human Genome Project was focused on the primary sequence. The epigenome is beyond that. I would like to argue that much of epigenetics is part of the "associated biological response" referred to in the definition of the exposome, essentially making epigenetics a feature of the exposome.

Work from the laboratory of Randy Jirtle was instrumental in showing that environmental factors could induce epigenetic modifications. In the widely cited example using the agouti gene, a gene that encodes coat color in mice, his laboratory showed that modifying the levels of methyl donors in the diet could completely change the coat color of a mouse via methylation of the agouti locus (Dolinoy, Weidman, Waterland, & Jirtle, 2006). While this was a visually powerful demonstration, it was a more powerful mechanistic finding. The genetic sequences of the mice were the same, but their appearance could be transformed through epigenetic modification, and more important, inherited. Subsequent work from the Jirtle laboratory, led by then graduate student and now department chair Dana Dolinoy, showed that the environmental contaminant bisphenol A caused hypomethylation of particular CpG islands, which could be reversed by enriching the diet in methyl donors (e.g., folic acid) (Dolinoy, Huang, & Jirtle, 2007). Moreover, work from the laboratory of Moshe Szyf demonstrated that the maternal grooming patterns of mice could impact the level of methylation in the offspring (Weaver, Meaney, & Szyf, 2006). Subsequent research has shown that child abuse and neglect and recreational drug use can confer epigenetic changes to one's genome. One should shudder at the possible epigenetic alterations that may occur upon exposure to more overtly toxic substances. There is evidence of major environmental exposures leading to transgenerational health effects that are likely due to such epigenetic changes.

Let us look at some of our Example children—Elmer and Eloise, two biological offspring of Ernest and Eliza. Both have inherited their genes from their parents. Yet there is a subset of these genes that are suppressed to allow preferential expression of either the mother's or father's DNA. Dozens of genes are thought to be regulated via imprinting. We assume that there is an equal chance of one or the other being expressed, but with these imprinted genes, either the father's or mother's gene is turned off primarily via methylation, throwing a wrench into our basic understanding of inheritance, since this occurs in a non-Mendelian fashion (Kappil, Lambertini, & Chen, 2015). This is important for several reasons, but from the exposome perspective, it is important to examine if these imprinted genes are more susceptible to environmental modification.

Another intriguing level of DNA regulation, which may have an impact on exposome research, is the chromatid end capping by telomeres.

Telomeres consist of repeats of a six nucleotide motif (TTAGGG) that acts to protect the chromosome from shortening during replication. Degradation and the resultant shortening of the telomeres are correlated with premature aging of the chromosome and an increased incidence of cancer and degenerative disorders. Elizabeth Blackburn, Carol Greider, and Jack Szostak shared the Nobel Prize in 2009 for the discovery of telomerase, the enzyme that adds the DNA-capping structures. It has also been posited that the telomere length may be an indicator of cumulative DNA damage, which could represent the exposome-induced associated biological responses (see Chapter 10, The Exposome in the Future) (Blackburn, Epel, & Lin, 2015). Elevated stress hormones, lack of sleep, depression, and chemical exposures have all been shown to shorten telomeres, which suggest that telomere length could become a component of an exposome index.

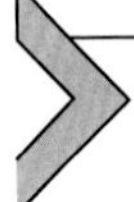

2.7 Using genetic know-how to better understand environmental influences

Investigators interested in the role of environmental factors in disease have benefited from the advances in the genetic toolbox, such as transgenic mice, mutant flies, and genetically modified worms. If a particular chemical is thought to interact with a particular gene or protein it is possible to simply order the strain of the fruit fly, *Caenorhabditis elegans*, or even mice that lack that particular gene to study that specific gene-environment interaction. Mutant lines of flies and worms can be generated in a short amount of time, and mouse strains that better represent genetic diversity, such as the Collaborative Cross and Diversity Outbred strains (www.jax.org), are now available to investigators. But it gets even better. Now scientists are collecting data on natural loss-of-function variants in the human population. Over time, we may have the ability to see how particular "human knockouts" alter proteomic and metabolic pathways, which would be a superb confirmation of studies conducted in animal models and may even obviate the need for some animal testing.

On the topic of impressive genetic advances, chalk another one up to the geneticists with the discovery and application of clustered regularly interspaced short palindromic repeats (CRISPR; Doudna & Sternberg, 2017). Building off of a bacterial defense system Jennifer Doudna,

Emmaneulle Charpentier, and Feng Zhang have set the stage for precision genetic manipulation in animals and humans. Investigators in the toxicology community are already using CRISPR screens to identify the genes responsible for the activation or deactivate of toxicants, proving yet another example of how advances in one field can lead to advances in another. The precision and relative ease of CRISPR technology may make our genome more malleable, but never more than our exposome.

I must reiterate discoveries in the field of genetics have been extraordinary. In fact, as scientists learn more about the regulation of DNA and studies based on deep sequencing or whole genome sequencing are conducted, we will likely find that genetics will unveil many more secrets. Indeed, many of the limitations noted in the chapter title may be resolved as the science progresses. In the future, our genes may explain even more than they do now, but our intuition tells us there is a limit and epigenetic changes will continue to play a role. It is important for us to learn what those limits are because this is where the interface with the environment will be critical. Everything beyond this hypothetical limit is by definition nongenetic- that is, exposome.

2.8 Obstacles and opportunities

It is important for exposome advocates to embrace the human genome. The tools, approaches, and strategies from the Human Genome Project and subsequent genetic endeavors will be extremely useful in understanding the impact of the environment on human health and disease. It must be a collaborative venture where genetics are used to provide a biological foundation. Environmental health scientists must continue to elevate the importance and awareness of the environmental influences on health without disparaging the genetic influences. It is not us versus them. Our genes do matter. We have and will continue to reap dividends from the study of our genome, but like many scientific endeavors that work their way into the public sphere, there has been a great deal of overpromising. The scientists and politicians over promise and the public over expects. The promises of the Human Genome Project have fallen short of expectations, but even if we had complete mastery of our genome, we would still be missing the external, exposome-like forces that will ultimately account for more than 50 percent of incidence for most chronic diseases.

The exposome has much to learn from the Human Genome Project. One of the most pertinent lessons is the need to have a clear goal. This is one of the major challenges currently facing exposome research. Sequencing the entire human genome was relatively straightforward. There was widespread agreement among scientists in the field that this was a worthwhile endeavor. Even so, there was considerable disagreement on the feasibility and approach. The exposome community must be prepared for a much greater level of skepticism and distraction, as it is more complex and there is not agreement on what it even is. Other lessons learned from the Human Genome Project that will be useful to the exposome community are addressed in the closing chapter.

2.9 Discussion questions

Companies such as 23andMe and Ancestory.com can provide extensive genetic information on an individual, and it is only a matter of time before personal genome sequencing is a reality. If cost were not an issue would you want to have your genome sequenced? Is there some information you would rather not know?

How do we protect this type of very invasive and private information? Should our healthcare and insurance providers have access to it?

How may having access to such information change one's behavior? For example, if you knew that you had a twofold increased risk of having a heart attack, would you alter your behavior?

References

Blackburn, E. H., Epel, E. S., & Lin, J. (2015). Human telomere biology: A contributory and interactive factor in aging, disease risks, and protection. *Science*, *350*(6265), 1193–1198. Available from https://doi.org/10.1126/science.aab3389.

Collins, F. S., Morgan, M., & Patrinos, A. (2003). The Human Genome Project: Lessons from large-scale biology. *Science*, *300*(5617), 286–290. Available from https://doi.org/10.1126/science.1084564.

Dolinoy, D. C., Huang, D., & Jirtle, R. L. (2007). Maternal nutrient supplementation counteracts bisphenol A-induced DNA hypomethylation in early development. *Proceedings of the National Academy of Sciences of the United States of America*, *104*(32), 13056–13061. Available from https://doi.org/10.1073/pnas.0703739104.

Dolinoy, D. C., Weidman, J. R., Waterland, R. A., & Jirtle, R. L. (2006). Maternal genistein alters coat color and protects Avy mouse offspring from obesity by modifying the fetal epigenome. *Environmental Health Perspectives*, *114*(4), 567–572. Available from https://doi.org/10.1289/ehp.8700.

Doudna, J. A., & Sternberg, S. H. (2017). *A crack in creation: Gene editing and the unthinkable power to control evolution*. Boston, MA: Houghton Mifflin Harcourt.

Falconer, D. S. (1960). *Introduction to quantitative genetics*. Edinburgh: Oliver and Boyd.

Franklin, R. E., & Gosling, R. G. (1953). Molecular configuration in sodium thymonucleate. *Nature*, *171*(4356), 740–741. Available from https://doi.org/10.1038/171740a0.

Janssens, A. C. J. W., & Joyner, M. J. (2019). Polygenic risk scores that predict common diseases using millions of single nucleotide polymorphisms: Is more, better? *Clinical Chemistry*, *65*(5), 609–611. Available from https://doi.org/10.1373/clinchem.2018.296103.

Kappil, M., Lambertini, L., & Chen, J. (2015). Environmental influences on genomic imprinting. *Current Environmental Health Reports*, *2*(2), 155–162. Available from https://doi.org/10.1007/s40572-015-0046-z.

Kim, U. J., Birren, B. W., Slepak, T., Mancino, V., Boysen, C., Kang, H. L., & Shizuya, H. (1996). Construction and characterization of a human bacterial artificial chromosome library. *Genomics*, *34*(2), 213–218. Available from https://doi.org/10.1006/geno.1996.0268.

Klein, C., Lohmann, K., & Ziegler, A. (2012). The promise and limitations of genome-wide association studies. *Journal of the American Medical Association*, *308*(18), 1867–1868.

Roberts, M. C., Khoury, M. J., & Mensah, G. A. (2019). Perspective: The clinical use of polygenic risk scores: Race, ethnicity, and health disparities. *Ethnicity and Disease*, *29*(3), 513–516. Available from https://doi.org/10.18865/ed.29.3.513.

Shreeve, J. (2004). *The genome war: How Craig Venter tried to capture the code of life and save the world/James Shreeve*. New York: Alfred A. Knopf.

Tenesa, A., & Haley, C. S. (2013). The heritability of human disease: Estimation, uses and abuses. *Nature Reviews Genetics*, *14*(2), 139–149. Available from https://doi.org/10.1038/nrg3377.

Watson, J. D., & Crick, F. H. (1953a). Genetical implications of the structure of deoxyribonucleic acid. *Nature*, *171*(4361), 964–967. Available from https://doi.org/10.1038/171964b0.

Watson, J. D., & Crick, F. H. (1953b). Molecular structure of nucleic acids; a structure for deoxyribose nucleic acid. *Nature*, *171*(4356), 737–738. Available from https://doi.org/10.1038/171737a0.

Weaver, I. C., Meaney, M. J., & Szyf, M. (2006). Maternal care effects on the hippocampal transcriptome and anxiety-mediated behaviors in the offspring that are reversible in adulthood. *Proceedings of the National Academy of Sciences of the United States of America*, *103*(9), 3480–3485. Available from https://doi.org/10.1073/pnas.0507526103.

Wilkins, M. H., Stokes, A. R., & Wilson, H. R. (1953). Molecular structure of deoxypentose nucleic acids. *Nature*, *171*(4356), 738–740. Available from https://doi.org/10.1038/171738a0.

Further reading

Mukherjee, S. (2016). *The gene: An intimate history* (First Scribner hardcover ed). New York: Scribner.

Quackenbush, J. (2011). *The human genome: The book of essential knowledge*. Watertown, MA: Charlesbridge.

Tam, V., Patel, N., Turcotte, M., Bosse, Y., Pare, G., & Meyre, D. (2019). Benefits and limitations of genome-wide association studies. *Nature Reviews Genetics*, *20*(8), 467–484. Available from https://doi.org/10.1038/s41576-019-0127-1.

CHAPTER THREE

Nurturing science

...variability from the indirect and direct action of the external conditions of life.

Charles Darwin

3.1 Introducing nurture

To counterbalance the genetic exploits described in Chapter 2, Genes, Genomes, Genomics: Advances, and Limitations, this chapter explores nongenetic influences commonly viewed under the guise of nurture. I open with a quote from the closing passages of *The Origin of Species* that addresses the environment or, to use Darwin's term, "the external conditions of life." The external conditions or surroundings work in concert with our genetic composition. The common phrase "nature versus nurture" implies an adversarial relationship, but it is better to view the two facets as complementary— even synergistic. For the vast majority of humans, their health is more a result of nongenetic factors, but it is certainly predicated upon their genetic makeup. This chapter discusses natural selection and human development in the context of the exposome.

Several years ago, Dean Jones and I had a series of philosophical discussions about the exposome. We both were comfortable with the uncertainty and complexity that we knew would come with this type of work. A biochemist by training, he had been working in the untargeted metabolomics space for several years. He, too, saw that the exposome represented much more than an exposure assessment challenge. We quickly agreed on the idea that the exposome was a fundamental part of biology and that when coupled with genetics had the potential to transform our knowledge of health and disease. This was when we started discussing the idea that the exposome could capture much of what we describe as nurture (Fig. 3.1). Together, nature and nurture reveal our phenotype— whether that be health status, physical traits, or disease symptoms. Nature- that is- genetics,

The Exposome.
DOI: https://doi.org/10.1016/B978-0-12-814079-6.00003-1

$$\underset{\text{Phenotype}}{\text{Health/disease}} = \underset{\text{Nature}}{\sum \text{Genetic factors}} + \underset{\text{Nurture}}{\sum \text{External factors}}$$

OR

$$\underset{\text{Nurture}}{\sum \text{External factors}} = \underset{\text{Phenotype}}{\text{Health/disease}} - \underset{\text{Nature}}{\sum \text{Genetic factors}}$$

Figure 3.1 Nature and nurture working together. Our phenotype, whether that be a physical trait or a state of health or disease, is a result of our nature or genetic programming plus our nurture or the external factors. By subtracting the nature/genetics from both sides of the equation, one can see that nurture can be described as the phenotype minus the nature or genetic component.

is firmly ensconced in our educational framework. From elementary school to university, there is a systematic way of studying and describing our genetic programming. Nature is biology and biology is science. Therefore, nature and genetics are part of science. Nurture, however, never seemed to get the same level of respect in biology classes. Nurture was often emphasized in psychology and sociology courses, and up until the past few decades, there was little biochemistry or molecular neuroscience related to the term. I sensed that the complexity of humanity and behavior was not nearly as quantifiable as genetics. Therefore, these concepts could not easily fit within the increasingly quantitative field of biology. I am not implying that the exposome can usurp the importance of psychology and sociology; rather, these external forces- from holding an infant, singing to a toddler, traumatic life experiences, or social network dynamics— all must be transferred into a biochemical signal (for example, altered cortisol or catecholamines). As such they open themselves up to biochemical measurements.

Thus, in the first edition of this book, I suggested that the exposome had the potential to provide a quantitative value for key aspects of nurture. If one accepts that the exposome represents the nongenetic drivers of a particular health condition, this seems reasonable. I am not ready to explore the exposome as a driver of evolution, as it runs on a timescale few can comprehend. Our health across our lifespan is a minuscule fraction in the course of human development. Each person contributes little to the evolution of the species as a whole. However, in each person's worldview, she or he contributes mightily to their own life, health, and satisfaction. Thus I would like to focus this chapter on the aspects of nurture that occur on the timescale of the average human lifespan.

The adversarial comparison between nature and nurture is unfortunate and often misinterpreted. A trait that a child exhibits that resembles their parents' traits is readily attributed to inheritance. The problem is that that does not always mean it was due to genetics. We learn and adopt behaviors from our caregivers. Facial expressions, voice intonations, and humor rely on interactions. Even more, physiological traits are often a combination of the two forces. Let us look at obesity as an example. There are certain genes that predispose many to obesity, but they cannot explain the sharp rise in obesity rates over the past 20 to 30 years. There has not been sufficient time for the genetic makeup of populations to explain this. The increase must come from nongenetic factors. There is also evidence that the eating habits of the mother can cause epigenetic changes to the developing child in utero, which represents a critical genetically carried environmental factor. The types of food we eat are based on cultural norms, availability, cost, the food industry, and personal decision-making. Of course, we have genetically programmed preferences for types of foods, but the primary drivers are nongenetic. They are part of our nurture, and to the degree that we can measure what we ingest, metabolize, and the resulting biological consequences, they fall under the rubric of the exposome.

We often think of nurture as tending to a seedling or caring for an infant, but nurture can be viewed as the entire constellation of factors that shape a being. The scope of the concept is vast and reveals the striking parallels between the concepts of nurture and the exposome. This is why it is critical to discuss the exposome within the context of nurture. I do not believe that the exposome can capture all of what nurture represents from a biochemical level, but it does attempt to capture more of it quantitatively than any other descriptor of nurture yet described.

We can't change our genes, but we can change our environment.

Linda Birnbaum, former director of the National Institute of Environmental Health Sciences (NIEHS).

This quote from the former director of NIEHS provides a good summary. CRISPR-Cas systems notwithstanding, changing one's genetics is still in the realm of science fiction. Once the sperm and egg unite, future alterations are primarily driven by the nongenetic forces. As I explore more in Chapter 9, The Exposome in the Community, we can change our environment. We can alter what we eat, how much we exercise, and our intake of recreational or pharmaceutical drugs. We can avoid dangerous activities and gravitate to healthier activities. All of these behaviors and activities, though, do not act within a vacuum. They all occur to and within our bodies that are derived from the genetic blueprint outlined in Chapter 2, Genes, Genomes, Genomics: Advances, and Limitations. Without input from the environment, including nutrition, social interactions, and shelter, we would not flourish. For the vast majority of the public, the environment plays a more important role to their health than does their genome, but as noted previously, it is predicated upon their genetic makeup. Therefore, an "either/or" argument is not at all informative. It is an interaction between the two. As seen in Fig. 3.2, the Example family can experience very different conditions throughout their days.

They work or go to school in different locations, they eat many meals that are different, and they are exposed to different physical conditions and populations of coworkers or fellow students. They are also subjected to different levels of physical and psychological stress. All of these nongenetic actions contribute to their nurture. Later, I will examine some of the key areas of nurture as they relate to exposome research. I do not attempt to

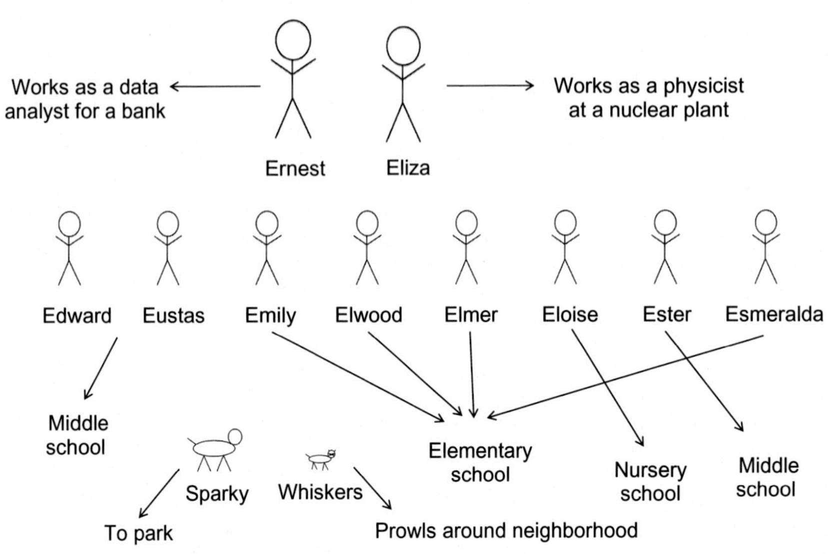

Figure 3.2 Families that live together have very different surroundings. One's residence does dictate a large part of their natural surroundings, but many of our waking hours are spent at jobs, in schools, or participating in various activities. These settings create very different conditions of nurture or exposures among the family members even though they have the same residence. Recall that Edward and Eustas are identical twins, Emily and Elwood are fraternal twins, and that Ester and Esmeralda are adopted.

examine all aspects of nurture, but rather focus on those that lend themselves to an exposome-based line of study. Specifically, it is the nongenetic forces have the most direct effects on the development of human disease. From my perspective the most important areas are nutrition, physical activity, psychosocial interactions, and exposure to pathogens.

3.2 Nutrition: feeding our exposome

Our bodies require energy for a variety of functions, and this energy comes in the form of food. We must consume food that can be converted into energy to fuel our biological systems. We also have several key nutrients that our bodies are incapable of making, and we must ingest these nutrients for proper function. However, there is a wide range of diets that can provide these essential nutrients. Think about all of the different types of diets around the world. Some cultures only consume plant-based diets; others are heavily carnivorous. Some are highly reliant on rice, others on corn. Yet, they all survive. This illustrates the range of foods that contain the key resources and our body's ability to adapt to the diet of our families and cultures. The human body is quite malleable. We have evolved to handle many different types of diets, but we must have a combination of protein, carbohydrate, and fat along with several key nutrients.

Specifically, we need 20 amino acids to make proteins; nine of these must come from our diet: histidine, isoleucine, leucine, lysine, methionine, phenylalanine, threonine, tryptophan, and valine. We can make the others from readily available precursors. Linoleic acid and alpha-linoleic acid are essential fatty acids we need from our diet. There are 13 vitamins that we must consume: vitamin A, vitamin C, vitamin D, vitamin E, vitamin K, B1 (thiamine), B2 (riboflavin), B3 (niacin), B5 (pantothenic acid), B6 (pyridoxine), B7 (biotin), B9 (folate), and B12 (cobalamin). We also need the minerals sodium, potassium, chloride, calcium, phosphorus, magnesium, iron, zinc, manganese, copper, chromium, iodine, molybdenum, selenium, and cobalt. Fortified diets in more developed nations have made malnutrition rare. But if a person eats an unhealthy diet, it is possible for various forms of malnutrition to occur. These levels of malnutrition may be more subtle than what was observed in the pre-industrial era, but they could have significant clinical manifestations. A well-balanced diet makes it relatively easy to consume the aforementioned essential components, at least to the point of avoiding the diseases known to occur from severe malnutrition, such as anemia, rickets, and scurvy from deficiencies of iron, vitamin D, or vitamin C, respectively. Unfortunately, our diets contain much more than the essential components. I discuss how our diet serves as a carrier system for thousands of synthetic chemicals in Chapter 4, The Environment: the Good, the Bad, and the Ugly.

3.3 Too much of a good thing

Thousands of years ago, food was scarce and it took a great deal of energy to acquire and prepare food. Eating was more intermittent in hunting-based structures or required much more energy as in the case of foragers. We ate larger meals at a much lower frequency or worked much harder for more frequent, yet smaller meals. There was a critical need to store energy for extended periods of time. Hominids evolved over millions of years, with *Homo sapiens* appearing over the past 200,000 years, and they did so under this intermittent food consumption model. Over the past century, due to modern farming practices and enhanced distribution systems, food is bountiful. In most communities in the United States, a grocery store is nearby. Convenience stores are common. We have refrigeration systems that allow us to keep weeks of food on hand. We have high-end restaurants and fast-food restaurants. Fifty years ago in the United States, the family meal was common. Families prepared meals in their kitchens. The use of preservatives was moderate. Restaurant meals were for special occasions. Now it is possible to obtain a prepared meal for 5 USD. Lifestyle and work schedules have forced more rapid food consumption. We eat more processed foods. We no longer have to search for our food. It takes very little effort and energy to obtain our nourishment.

It will be interesting to see the impact of the industrial revolution and modernity on human evolution. There is a strange irony here. Over the past 100 years, human lifespan has doubled. Does this mean that our altered eating habits are a good thing? Could that be part of the increased lifespan? Improved quality of diet has likely contributed to our increased lifespan, but it is difficult to disentangle these effects from the improvements in water quality and distribution, sanitation, and vaccines. There has not been a major impact on the span of the reproductive window that allows the passage of genetic material at a species level. Yes, men can father children at older ages, and women tend to have children at older ages, but this has more to do with societal norms and access to reproductive care than a shift in reproductive fitness. Regardless, in developed countries over the past century, our diet has not had a major impact on our population dynamics.

In modern society, there has been a frightening increase in the rates of obesity and Type 2 diabetes. We are eating too much. A substantial portion of those who are overweight recognize the need to decrease their weight, and we are bombarded with new fad diets every few years. Given the general public's poor understanding of nutritional biochemistry, it can be easy to be swayed by a commercial or a colleague promoting a new diet. People are searching for a diet that can let them feel the satisfaction to which they have grown accustomed, but with a net decrease in calories. The set point for satiety has shifted dramatically with the increased sugar and salt content in our food, which makes it very difficult for people to shift to a higher quality of foods. As we increase our caloric intake above what is required, the excess comes at a hefty price. In addition to weight gain, we are also exposed to more of the chemicals found in our foods (Fig. 3.3).

As our eating habits evolved and we started fortifying foods, we have had incredible public health improvements. But like what we have seen with chemical exposures, as we address major problems, we are left with a myriad of minor problems that when put together remain a major problem.

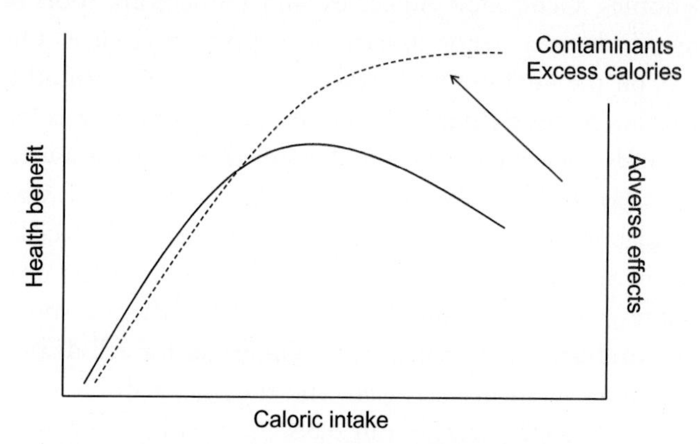

Figure 3.3 Nutrient needs, dietary excesses. Deficiencies in specific nutrients can cause a variety of health conditions and diseases. Once we get to the point of satisfying our biochemical needs, additional levels of nutrients and calories do not provide additional benefit. For many nutrients, especially the lipophilic vitamins, can accumulate in the body and lead to toxicity. Optimizing intake of nutrients can minimize adverse exposures.

For example, we have reduced the levels of exposure to heavy metals such as lead and mercury, which are major advances for public health, but now we are dealing with the effects of many small cuts. That is to say the levels of highly toxic chemicals have decreased, but we are exposed to *many more* chemicals that individually have a lower, but complex toxicity profile. When combined, there may be additive and even synergistic toxic effects. As it turns out, many of these chemicals are making their way into our bodies through our diets. If we are exposed to 10 times or 100 times more chemicals than we were a century ago, we must start asking about their aggregate and combined effects. Unfortunately, this is not something we have been capable of doing from an experimentalor regulatory standpoint.

3.4 Dietary insight from the laboratory

Nutritional studies have an advantage in that we can collect a large amount of data from the human population. We can take dietary records in patient populations, study food intake patterns across cultures, and have data from various world events (such as the Great Depression in the United States or the Dutch Hunger Winter). Yet, we can also go to the other end of the spectrum in the laboratory. As noted in Chapter 2, Genes, Genomes, Genomics: Advances, and Limitations, tools of geneticists promise to have a major impact on exposome science. One of the earlier forays on the genome front was the use of a small roundworm as a model organism. Sydney Brenner was a top geneticist and was looking for a system to study inheritance and DNA regulation. He was aware that the roundworm *Caenorhabditis elegans* (*C. elegans*) was one of the simplest and most manipulatable animals. They have fewer than 1,000 cells in their body and only 302 neurons. They reproduce asexually (they are primarily hermaphrodites). Their biological simplicity makes them a superb model in which to conduct mutational screens. The genes for cell death (apoptosis) were first discovered in *C. elegans*, and they have also revealed critical insights into the linkage between our diets and health.

Cynthia Kenyon, now at the company Calico and the University of California at San Francisco, completed her postdoctoral studies with Brenner at the University of Cambridge and then embarked on a career focused on the genetic determinants of aging using *C. elegans*. Her group discovered genes that controlled longevity in the roundworm (Kenyon, Chang, Gensch, Rudner, & Tabtiang, 1993). As it turned out, the genetic pathways that controlled lifespan in the roundworm were part of the insulin signaling pathway connecting them to nutrition and energetics. Given insulin's critical role in glucose and energy regulation, it became clear that food ingestion was key. This discovery helped spawn an explosion of research examining the role of diet and energy balance in maintaining a healthy lifespan.

The concepts of intermittent fasting and diets with a low glycemic index are, in part, based upon this work. Although one could have modeled this after our ancestral eating patterns, it was the research in *C. elegans* that provided the biological foundation for this line of work. Humans are much more complex and sophisticated than worms, but our metabolism is not as different as one may imagine. Kenyon also walks the walk, in which she has based her diet upon the data from the roundworm and other experimental data on foods that alter blood glucose (Lee, Murphy, & Kenyon, 2009). My laboratory has embarked on a collaboration with Coleen Murphy at Princeton, who trained with Kenyon and discovered some of the regulatory genes involved in the life extension (Murphy et al., 2003), to examine metabolism in *C. elegans*. Our long-term goal is to employ *C. elegans* to study the exposome in a manner analogous to how the model was used to study the genome. It has been encouraging to have the support of the *C. elegans* community for our exposome efforts given their firm grounding in genetics.

Caloric restriction can extend life in other models, such as mice and rats, and this has been known for decades (Holzenberger et al., 2003; Masoro, 2005). There has not been a sufficient amount of time to conduct lifespan studies in humans, but there is evidence that caloric restriction may slow aging processes (as discussed later). Restricting calories tends to increase lifespan and the overall health of the organism. As noted above, it has become too easy to eat large amounts of calories. This has a direct impact on our exposome. If we exceed our nutritional needs, excess food comes with little benefit and additional chemicals (see Fig. 3.3). Nutrition is part of nurture, and the exposome must fully embrace diet and nutrition at all levels.

3.5 Physical activity—exercising your exposome

I studied exercise physiology as an undergraduate and was enamored with human performance. How was it that humans could perform incredible feats? How did muscles respond to training in such a way as to continually increase in strength? How does the heart adapt to aerobic exercise to pump more blood and deliver more oxygen to the tissues? These are some of the questions to which I wanted answers. I had the pleasure of working with the late Mel Williams, who conducted some of the first studies demonstrating that the infusion of one's stored red blood cells increased the maximal oxygen uptake and enhanced human performance—that is—blood doping, as well as Richard Kreider, who has emerged as a key expert in sports nutrition. Much of this work was conducted in high-performing athletes where one can observe the effects of intense training on performance. My interest in high-performing athletes eventually waned as I realized that moderate physical activity in the general population was a much more important topic of study from a public health perspective.

Modernity has done much to decrease our reliance on activity. Although mobility scooters and electric wheelchairs are essential for those with disabilities, recreational scooters, Segways, and reliance on cars for short trips have reduced our general activity level to our physical detriment. Far too often in the U.S., many cities and communities are not designed to make it easier to walk to schools and stores. Over the past 40 to 50 years, the number of children using self-propelled forms of transport (walking, biking) to school has decreased dramatically. The trends are self-evident, but there is very little effort put into changing these trends. I revisit this in Chapter 9, The Exposome in the Community, but I believe that increased physical activity is one of the most important changes needed in modern society. Commercial gyms, recreational sports, and exercise equipment are all good, but the key issue is that our activity must be increased at a societal level. Our day-to-day lives do not require as much physical activity as we used to need to survive. Our activity and labor have been reduced as routine activities, such as obtaining fuel for warming our homes to collecting food for meals, have become more automated or otherwise simplified. More jobs and professions can be executed from an office and computer terminal.

Physical activity is more than cardiovascular fitness. There are strong links between inactivity and increased rates of heart disease. Moderate aerobic exercise increases heart efficiency, oxygen transfer from hemoglobin, and increased arborization of small blood vessels. These are all positive gains in our physiology. Our muscles also benefit from exercise. Activities that put strain on muscles, such as lifting, gripping, and moving, activate neuronal circuits. Thus, the muscles themselves benefit, but so do the neuronal circuits in the brain. The term plasticity describes the ability of the brain or other components of the nervous system to be altered due to some sort of input. We tend to think of plastic objects as being hard, but in fact, when plastics were brought into the industrial process it was for their high level of malleability. Plastics could be made into all types of shapes. Thus, brain plasticity reflects how much our neurons can be altered by the environment. Data over the past decade have revealed that increased physical activity can enhance neuronal function in the brain. The work of Fred Gage at the Salk Institute has illustrated that our prior view that new neurons cannot continue to be formed was wrong and physical activity can promote neurogenesis in the brain (Gage, 2019; Kuhn, Toda, & Gage, 2018). We tend to think that our brain tells our body to move in a unidirectional manner, but the movement itself sends signals back to the brain and these signals are essential for the development and maintenance of the brain.

Although the net benefits of exercise are nearly always positive, there are issues with physical activity within the context of the exposome. Exercising increases respiration and metabolism and puts physical stress on our joints, bones, ligaments, tendons, and muscles. If one is exercising next to a highway or in a polluted area, it will increase one's exposure to the ambient toxicants. At what point does exercising in these conditions negate the effects of the exercise itself? Except for those with known susceptibilities—for example, asthmatics—the net benefit of physical activity tends to outweigh the adverse effects. When smog or other air quality alerts are issued, it is probably best to avoid outdoor exercise at that time. People tend to be very creative in finding ways not to exercise, so warnings about not exercising at certain times of day without also providing information on safer times or other alternatives may lead to decreased activity. For example, if a person is only able to exercise in the afternoon due to her work schedule, discouraging exercise in the afternoon may have the net effect of her not exercising at all. When in fact the primary concern is exercising *outside* when the ozone, temperature, or particulate matter are high, while *indoor* exercise can avoid most of these risks. Preferably, we should direct people to the healthiest forms of exercise and the best conditions as described in Chapter 9, The Exposome in the Community, on the modifiable exposome. The Example family will also experience very different exposures from their general recreation and hobbies. Workplace and home exposures to pesticides, solvents, and pathogens can pose significant health risks. In Fig. 3.4 we can see that

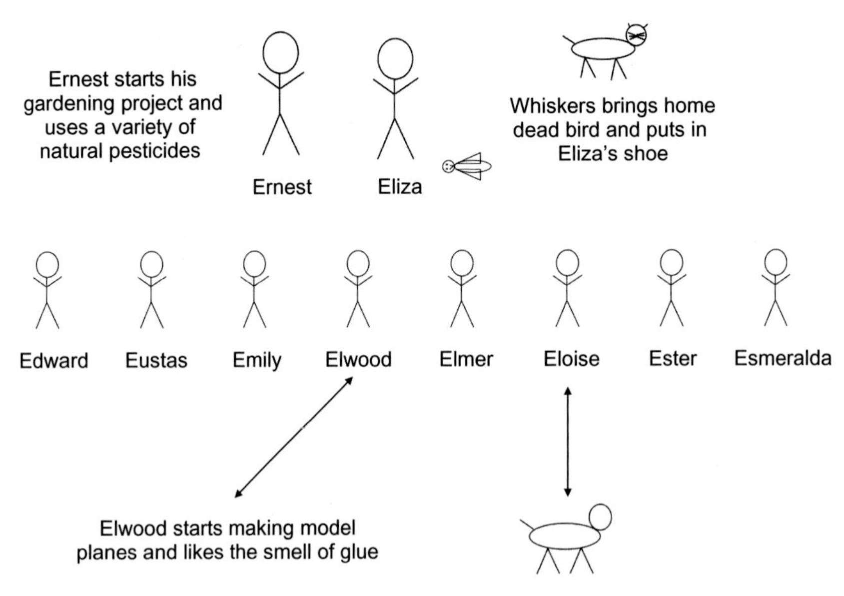

Figure 3.4 Diverse interests, diverse exposures. In addition to the conditions of nurture that are found among a family unit, family members can participate in a diverse array of activities that can introduce influential effects. Chemicals used to control fleas or pests on plants can find their way back into the home, as can a wide range of insects. Recreational use or misuse of chemicals represents a major exposure pathway.

different family members have different exposures from the ways they interact with their surroundings. This further illustrates the challenge of merely relying on one's physical place of residence to estimate overall exposures.

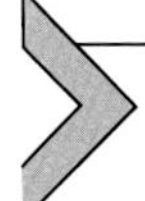

3.6 Psychosocial interactions and cognitive development: your humanity and the exposome

The human brain takes many years to develop and mature. While our five senses are somewhat developed after birth, our language, recognition of self, and significant problem-solving skills develop much later and are greatly influenced by our surroundings. Infants must engage with adults to see objects and activities, hear words and intonations, feel touch and grasp, smell a broad variety of odors, and taste milk or formula, and eventually foods. This engagement with the environment provides the possibility that these senses and skills will properly develop. Conversely, neglect, poor nutrition, and abuse can leave lasting adverse effects on the child resulting in deficiencies.

In our Example family, we know that the six biological children had very similar early life experiences, but Ester and Esmerelda, who were adopted, spent a period of time that was much different. Ester was malnourished for a few months and Esmerelda suffered from early childhood trauma. They both lived in different areas of the world and had very different environmental exposures. It is highly likely that both of them have emotional and physiological consequences that may be carried via epigenetic changes (Barboza Solis et al., 2015). Thus it is more than the inherited genetics that makes the Examples' biological children different from their adopted children. Their early development had a major influence on their being, and any adverse effects may have lasting consequences. For example, traumatic experiences that occur early in life have been associated with epigenetic changes in brain pathways involved in mood and affect (Cecil et al., 2016).

The way we interact with the world is changing. The shift from agrarian societies to more urbanized societies changes social structures. Our social networks are changing in scale and quality. The term "social networks" evokes the thoughts of Facebook, Instagram, Snapchat, and Twitter, but I use it to describe the dynamics of human interaction. Ironically, with our increasing population, there has been an increase in the incidence of social isolation and loneliness. There are distinct psychological and physiological consequences of loneliness, and it is more common in older people. Since we are living much longer, there is a growing need to address loneliness on a population scale. This is yet another example of the importance of sociologists, psychologists, psychiatrists, and other behavioral health professionals in exposome research. It is interesting to consider how electronic social media platforms have changed social dynamics.

Recent data have demonstrated adverse effects of social media platforms such as Instagram, Facebook, and Snapchat, as they force unrealistic expectations and comparisons. But our social networks are much more than social apps. Our actual social networks have critical effects on our health, providing comfort during stressful times, preventing loneliness, and creating a safety net of support for life in general. A drive to understand complex social networks was a major contributor to the development of the field of network science (Watts, 2003), which is covered in Chapter 7, Pathways and Networks. How we interact with our friends, families, and communities has a significant impact on our well-being.

The complexities of social interactions impact our physiological systems and represent a major part of our nurture. There are positive and negative ways of dealing with challenges. Raising one's readiness to deal with a crisis but not having strategies that allow one to rise to that challenge, may represent a psychological over- idling (as discussed later). In general, being anxious about a project or presentation can range from mild worry to notable fear. After the event is over, there is typically a sense of relief and accomplishment. When one is consumed by issues that are out of their control, there is unlikely to be any sort of resolution or release. How people respond to the range of challenges that fall within and outside of their control represents important physiological and psychological stressors and factors that should be captured by exposome research.

Now that we have reviewed some of the key contributors to our nurturing, we will explore the biological evidence of these contributors. Is it possible to measure the positive and negative aspects of nurture? At the risk of being simplistic, there are a few types of approaches that do just this.

3.7 What actions can the Example family take to improve their exposome?

The Example family had DNA analysis performed by a company. How will this change their behavior? Even if the tests reveal a risk allele for a gene or two, at this time, there are no genetic interventions available. Since the FDA only allows reporting of genes that reveal risk, it is highly unlikely that the medical community would support any genetic intervention. What it can suggest are behaviors and lifestyle changes that are known to influence the outcome of that disease. Even with that information, it is also probable that all of the suggested changes to behavior will fall under the heading of the exposome.

For example, mother Eliza has an increased genetic risk of heart disease. She eats a healthy diet and exercises on a regular basis, so there is not much more she can do at this point. Of course, the test results provide useful information for her doctor, but she had already reported that one of her parents died at an early age of heart disease. Thus the genetic testing has not impacted her behaviors or health care. Even so, any changes that would result are still part of the exposome. Ernest has a risk allele for macular degeneration. He will be sure to visit his ophthalmologist on an annual basis, but there is not much he can do on a day-to-day basis except to wear sunglasses and limit the time he spends staring at a computer or phone screen. Again, the genetic testing provided information, but most of the solutions are exclusively about altering our nurture.

For most of the year the Example family splits up daily as they attend school and work, but they share their environment at home (see Fig. 3.2). Which site has the biggest influence on their health? Many investigators believe that knowing a person's address can provide information on health. Recent data indicate that it may be workplace and previous exposures that are more important than home address (Chung, Kannan, Louis, & Patel, 2018), providing further evidence of the major impact of our surroundings that nurture our well-being. Hobbies can provide important developmental input by activating brain pathways, strengthening muscle groups, and providing psychological benefits by reducing stress.

It turns out that 50 percent of the Example family members have a genetic inability to break down lactose (lactose intolerance). If the entire family takes a trip to a restaurant for ice cream, half of the members of the Example family will have an adverse reaction. The health effect is a combination of genetic vulnerability and exposome input. The parents are aware of the adverse reactions that some of their children have to dairy products, so they are presented with a fairly straightforward solution—alter the exposome of those children be restricting their intake of lactose-containing foods (Fig. 3.5).

Ernest took up gardening a few years ago. He viewed it as a way to spend time outdoors and provide fresh vegetables for his family. Ernest thought it was best not to use synthetic pesticides in their garden and opted for an organic route. He uses a *Chrysanthemum* tea, a chili pepper concoction, some natural pyrethrums, and rotenone—all of which he acquired from an organic gardening store. He wore gloves when he applied the products. He was not aware that the chemicals, albeit natural, are nearly identical to those used in synthetic pesticides. This illustrates an important point. We tend to view synthetic chemicals as being worse than natural products. Certainly, exposures to unnecessary synthetic chemicals are bad, but plants and animals have evolved to synthesize a myriad

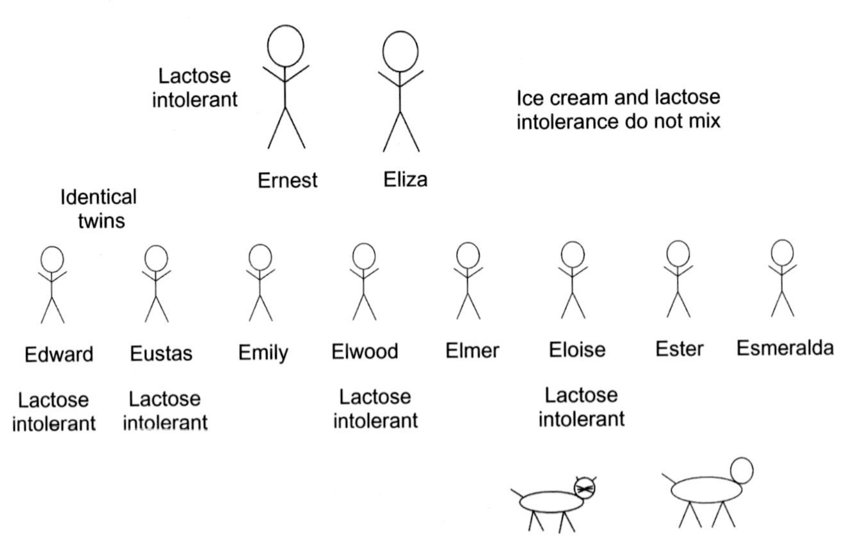

Figure 3.5 Lactose intolerance illustrates how nurture reveals phenotype. In this example the family goes out for ice cream. The family members with the genetic predisposition for lactose intolerance have gastrointestinal distress, while the others do not. It is the combination of the genetic trait and the nurture or environment that reveals the biological trait. Although we do not know the lactose intolerance state of Whiskers and Sparky, dogs and cats can suffer from the same enzyme deficiency as humans.

of chemicals to protect themselves from predators. These external chemicals, whether derived from nature or a laboratory, have the potential to alter our health. We cannot merely view the natural label as being safe. Thus organic gardening that avoids the use of chemical pesticides is good, but only if the synthetic chemicals are not replaced with dangerous natural chemicals. Our understanding of the externalities that affect our health is lacking, and exposome research is positioned to improve this.

3.8 Immunity and the exposome

There has not been much research on how the exposome impacts our immune system, but as noted above, it is a critical mechanism of sensing our exposome. At the risk of stepping on the toes of the field of infectious disease, I believe that the parasites, bacteria, viruses, and other pathogens that threaten our health are part of the exposome. I am not suggesting that we relegate infectious disease to an exposome subdiscipline, but rather that these pathogens represent some of the most important aspects of the exposome. Furthermore, more than half of human infectious diseases are zoonotic, meaning that our interactions with animal species are critical to understand human health (https://wwwn.cdc.gov/norsdashboard/). This is especially true in regions of the world where there is poor control of disease vectors (Gebreyes et al., 2014). A wide variety of pesticides are used to control mosquitoes, fleas, and ticks and that we must balance the risks and benefits of vector control. It is critical to use the lowest level of chemicals possible to achieve vector control and minimize human exposure to chemicals.

It is easy to imagine a synthetic chemical as being foreign to our body, but there are many other foreign compounds that influence our health. For example, we have not discussed Sparky and Whiskers. They are more than cute illustrated diversions. There is compelling evidence that having animals in the home can impact vulnerability to allergies. When one thinks of pet hair and dander, and the other things the pets track into the home, it is reasonable to suspect that their presence would increase the rates of allergies and asthma. This is not the case at all. In fact, the opposite is true. The presence of pets in the home decreases the likelihood of asthma and allergy and does not appear to have a negative impact on respiratory immune function. How can this be? Our immune systems are trained/primed by our exposures to allergens and other foreign molecules. The immune system is designed to recognize self from nonself. The identification of foreign molecules is crucial for our survival. The molecules that are recognized by the immune system are referred to as antigens and the specific sites on which antibodies recognize antigens are called epitopes. It appears that having a wide variety of exposure to antigens decreases the probability of developing allergies. In a way the immune system is a key sensor of our exposome. This is another important example of how our environment shapes who we are.

Recall that the discussion of health in Chapter 1, The Exposome: Purpose, Definitions, and Scope, referred to resilience. One of the most important aspects of resilience is the ability of the immune system to fight off infection. Furthermore, infections can increase vulnerability to other environmental challenges. A gastrointestinal virus can break down the intestinal barrier, making it easier for food-borne chemicals to enter the bloodstream. A respiratory virus can make one more vulnerable to particulate matter and exacerbate underlying lung dysfunction, such as asthma.

There is also strong evidence that exposure to chemicals, such as heavy metals, can suppress immune function, increasing vulnerability to pathogens (Chandler et al., 2019). Thus, our immune system sits at the interface of nature and nurture. We have genetically programmed immune systems optimized to respond to an unlimited number of antigens, but it is the exposures-that is-nurture that provides the antigens to develop and prime our immune systems.

One of the most important physiological systems for regulating our response to stress is our neuroendocrine system. The hypothalamic-pituitary-axis serves as the master control of our neuroendocrine system. The hypothalamus provides input to the anterior and posterior pituitary. The anterior pituitary releases several regulatory hormones such as growth hormone, prolactin, thyroid-stimulating hormone, melatonin-stimulating hormone, adrenocorticotropic hormone (ACTH), follicle-stimulating hormone, and luteinizing hormone. These hormones are capable of activating dozens of key pathways that regulate our response to our changing environment. ACTH sends signals to the adrenal glands, which releases cortisol, aldosterone, and adrenaline (epinephrine). Cortisol is considered a key stress hormone and is thus measured as part of allostatic load. Aldosterone is critical for fluid homeostasis and adrenaline is essential for liberating energy stores to allow rapid and high-intensity muscle activity. The hypothalamic-pituitary-adrenal axis is at the core of our response to external stressors. It serves as a key regulator to Darwin's "external forces of life." As such, exposome researchers must incorporate these systems in their analyses as they are critical to how our bodies response to nurture, à la the exposome.

3.9 Nurture and evolution

As we examine many of the components of nurture, it is helpful to review how they contribute to evolution. Evolution operates on a timescale that is difficult for most humans to grasp, because the evolution of species occurs over thousands and millions of years, and thus is not witnessed by humans who tend to live for less than a century. Except for the evolutionary biologist, the general citizen of the world is thinking of a self-centered 10 to 50 year timeline. Even with a focus on public and population health, we are still primarily concerned with the fitness of the individual for the sake of that individual, not for their evolutionary gene-spreading prowess. This is not to say that we do not see a bit of natural selection, as witnessed by the annual Darwin Awards (awards go to unfortunate individuals who have removed themselves from the gene pool due to poor decision-making). In the concluding chapter of *Origin of Species*, Darwin cites the critical role of environmental variability in natural selection, and specifically he refers to the "variability from the indirect and direct action of the external conditions of life." Although the words environment and exposome are absent, the actions of the external conditions of life are essential to the argument. Nearly 200 years later those external conditions of life are now subject to systematic examination under the rubric of the exposome.

3.10 Evaluating nurture

How does one measure the quality of nurture? Even if one limits the discussion to the aforementioned areas, it is still a challenging task. We do have many ways of measuring the current health state of an individual. We can interrogate their genome. The components of nurture or the exposome are what connect the genome to the phenotype. Thus reliable health endpoints provide a good starting point. Although crude, our lifespan is a strong indicator of a healthy exposome. Given the relatively small changes in the human genome over the past 100 years, the near doubling of human lifespan in that time is a testament to our improved exposome. Some of the most critical advancements noted above include clean water, improved sanitation, and vaccines. All of these decrease our exposure to infectious agents. With the dramatic reduction in infectious disease, we now find the human population having to deal with chronic disease conditions. These chronic conditions have a major age-related component, such that increasing age increases the incidence and severity of the disease. Thus we are now much more concerned about our parts wearing out (knee or hip replacements, liver cirrhosis, and muscle loss). To introduce an automobile analogy, the longer you keep a car, the more you witness parts wearing out and needing replacement. By examining the trajectory of age-related health parameters, it is possible to track the rates of decline as a means of measuring wear and tear.

Evaluating the wear and tear on our bodies provided the motivation for the development of the concept of allostatic load. Homeostatic mechanisms are found throughout our cells and organs. Our bodies must respond and adapt to change. Homeostasis describes a process where the system returns to the initial set point. Allostasis represents how the system adapts to those changes. McEwen et al. introduced the concept of allostatic load as a way to measure the ongoing wear and tear on the human body (McEwen, 1998; Seeman, McEwen, Rowe, & Singer, 2001). Their conceptualization of allostatic load was based upon measures that could be collected in the clinical setting. The measures capture altered stress dynamics and the physiological features that reflect enhanced activation of these stress pathways.

Components of allostatic load

Systolic blood pressure
Diastolic blood pressure
Waist-to-hip ratio
Urinary cortisol
Urinary norepinephrine
Urinary epinephrine
Glycosylated hemoglobin
HDL cholesterol
Serum dihydroepiandrosterone sulfate

Blood pressure values provide a broad assessment of cardiovascular health. Waist-to-hip ratio provides information on fat deposition and net caloric intake. Serum cholesterol is an excellent indicator of atherosclerosis risk, with elevated HDL being protective. Glycosylated hemoglobin provides information on steady-state glucose metabolism. Serum dihydroepiandrosterone sulfate (DHEA-S) provides information on the level of hypothalamic-pituitary-adrenal access function, as does overnight urinary cortisol. Overnight urinary epinephrine (adrenaline) and norepinephrine (noradrenaline) provide information on the level of sympathetic nervous system activity. A population can then be ranked on these measures, and those people that fall into the 25th percentile of each measure (top 25th percentile for blood pressure, waist-to-hip ratio, total cholesterol, glycosylated hemoglobin, urinary cortisol, norepinephrine, and epinephrine, or bottom 25th percentile for HDL cholesterol and DHEA-S) were shown to have worse health outcomes (Seeman et al., 2001). The values derived from the allostatic load based on these relatively simple tests are predictive of lifespan, which is remarkable (Karlamangla, Singer, & Seeman, 2006).

There has been a great deal of effort into developing and validating the concept of allostatic load, and I am grateful to the efforts of the investigators who did this. That said, I believe that the allostatic load concept needs an update, both from a conceptual and experimental standpoint. As noted above, we have many new techniques that detect wear and tear on our bodies. Many of these outputs are more sensitive than the rather coarse clinical measures currently used to define allostatic load. These clinical indices require a substantial amount of time by trained personnel. Of course, the challenge is capturing the clinical measures of the allostatic load, which requires overnight urine, plasma, and other physical exam parameters.

Perhaps there is room for an allostatic load 2.0. From an exposome perspective, the integration of omic-related methodologies would be extremely helpful to determine if the underlying biochemical foundation of allostatic load could be measured in biological fluids. I have thought for several years that there must be an allostatic load signature embedded within our metabolomic network, and I introduced this idea at a workshop at NIEHS (Dennis et al., 2016). That is to say if high-resolution mass spectrometry—based metabolomics (like that described in Chapter 5: Measuring Exposures and their Impacts: Practical and Analytical) were compared against a continuum of allostatic load measures one could identify features that were strongly correlated to allostatic load. Analysis of DNA and protein adducts could also be useful by providing evidence of cellular damage from environmental exposures. This would likely require a large number of subjects, but is not out of line with the numbers being studied in ongoing exposome research. If a surrogate of allostatic load could be derived from plasma samples it could be readily applied to studies that have stored samples.

The classical fight-or-flight/rest-and-digest concept is important here. Our stress response system has evolved for survival under immediate threats. David Goldstein at the National Institute of Neurological Disease and Stroke likes to use the analogy of a getaway car (Goldstein, 2006). If a person is robbing a bank it is prudent to leave the car idling so that a quick escape is possible. For the vehicle, it is not good to be idling as this low-level activity causes wear and tear on the engine. It is using fuel and wearing parts, but not making any forward progress or transporting anything. But the seconds it takes to restart the car could be the difference between escaping with the cash and getting caught by police. Thus, the would-be bank robber is willing to sacrifice the long-term health of the vehicle for the immediate benefit of escaping with the money. Our bodies have developed similarly. We must always be ready to jump out of the way of a speeding car, dodge a falling tree limb, or flee from a threatening animal. Yet, our idling systems wear down our parts.

When a person says that they are stressed out, this is likely quite accurate. They may well have a prolonged upregulation of their stress hormones. The lack of sleep, poor nutrition, and reduced exercise that commonly follow demanding deadlines or other external pressures put a person in a constant low idle. Of course, it is okay to shut off a car, but our bodies, even when asleep, must be idling a bit. However, we should strive to keep that idle low when it should be low. Back to Goldstein's car analogy—in the prefuel injection days, carburetors were needed to mix the fuel and air to facilitate combustion. Mechanics used to encourage driving one's car at high speeds to "blow out the carbon," meaning by accelerating the car and driving it near its maximum rate one can remove deposits in the engine. Whether or not this practice worked or was just an excuse to drive fast is not clear, but conceptually it is intriguing. Do our bodies prefer low-level stressors all of the time or intermittent higher stressors?

3.11 Tick-tock, tick-tock. Is that my epigenetic clock?

Steve Horvath at UCLA has proposed that a subset of methylation sites can predict the relative age of an individual (Horvath, 2013). Using data from genome-wide methylation studies, he created a series of algorithms that estimate the epigenetic age a person and different cell types. By examining what a normal epigenome is over the life course, Horvath plotted age versus epigenetic changes. Thus given a person's chronical age, she or he should have an epigenome that looks like a person of that age. If, however, a person's epigenome looked like that of an older or younger individual, one could conclude that that person has accelerated or decelerated aging. The idea is that one can compare one's chronological age (based upon date of birth) with one's epigenetic age. This may be why physical exercise is universally beneficial to human health. Vigorous walking, running, biking, and strength training increase our metabolic rates, blood pressure, and catecholamine levels, but they do so in a controlled way. These activities initiate a series of biological processes that restore one's capacity to repeat the behavior at a later time, and many of these regulatory changes that lead to prolonged changes in biological processes are epigenetic in nature. There is evidence that overly intense or prolonged exercise can be detrimental to health, but this is only at the extremes. Similarly, the capping of our chromosomes by telomeres may provide an index of chromosomal wear and tear. As people age, their telomeres shorten. Since it is age-associated, one can examine telomere length and use it as a surrogate for aging (Epel et al., 2004).

Let us go back to the Example family. The father, Ernest, realizes that he spends too much time sitting at his desk, and his son Edward lost interest in the sports team he was on and has become more sedentary. Together, they start a program of moderately intense physical activity to reduce stress and improve fitness. The other family members continue their normal activities (and unknowingly serve as a control group). A couple of times each week the father and son expend energy, cause microtears in their muscles, and put pressure on their joints. Their catecholamine levels increase and their cardiac output increases. This is a controlled activity, not slow idling, and it is a full activation of their systems to perform physical activity. As expected, their bodies respond very well to this. Their muscles rebuild, their energy stores are repleted, and their cardiac output returns to normal. Over time, we see several positive changes from participation in physical activity even though the actual exercise bouts can be viewed as stressful from a physical and psychological perspective. Their resting heart rates decrease due to the increased output from their hearts for each beat and their muscles rebuild slightly stronger. The result is an ability to perform their activity at a higher level and for a longer time than when they started. Would all members of the Example family have adapted in a similar manner? Although all would derive some benefit, one's response to exercise differs based on their genetic profile and other nongenetic factors, such as nutrition and lifestyle (Fig. 3.6).

Before the industrial revolution, the average lifespan of an individual was 30 to 40 years. Injury, infection, and malnutrition were some of the biggest drivers of death. As we moved into the post-industrial era, many of these threats diminished. Public health interventions have extended our lifespans. We generally have fewer events of high stress, and by that I mean physical external threats that cause massive releases of catecholamines

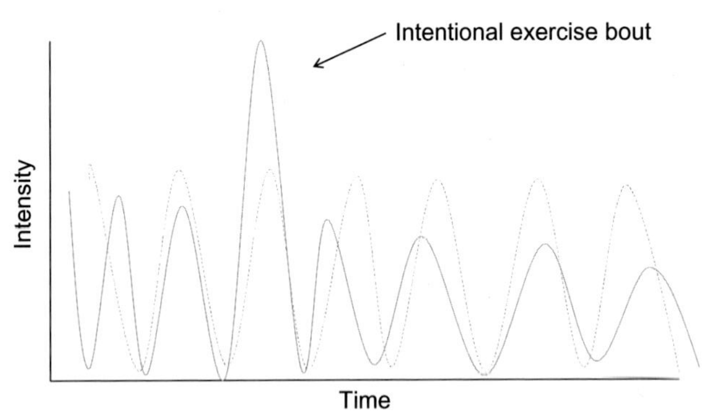

Figure 3.6 The ups and downs of stress. The y-axis shows the intensity of the stressful activity or experience and the x-axis shows time.The term stress it generally used to describe negative effects; however, stress is necessary to challenge our system and enhance performance. This is a critical aspect of nurture. In this example, the intensity of stress could be measured by catecholamine levels, blood pressure, oxygen consumption, or cortisol release. These external forces provide stressors that activate biological pathways that can lead to strong and more robust systems. At the same time, low-level, repeated, and chronic stressors are more likely to wear out biological systems. The dotted line shows a person who is subjected to frequent and repeated bouts of stress. The solid lines show a person who has a controlled bout of exercise. Over time the controlled higher peaks help reduce the everyday baseline. Heart rate is an excellent example. Even though exercise increases heart rate for a given period over a person's life, their heart beats less if they exercise regularly because of the adaptations that impact every hour of their life.

(adrenaline and noradrenaline, a.k.a. epinephrine and norepinephrine—the catecholamines) needed for survival; however, modern-day life has created a series of low-grade stressors that cause low-grade catecholamine release. Thus we have managed to find ways to increase our idling rate that are not necessary needed for our survival. Low-grade continual stress wears down our parts over extended periods of time in an almost imperceptible way and sets up the circumstances for a multitude of chronic health conditions. This ongoing low-level stress combined with our increased lifespan is manifested in increased rates of cardiovascular disease, anxiety, and depression.

There are exciting examples of how to measure aspects of our historical nurture. Manish Arora at Mt. Sinai has pioneered the use of laser ablation coupled to mass spectrometry to reconstruct past exposures (Morishita & Arora, 2017). This work can be performed on a person's deciduous (baby) teeth—making sample collection noninvasive. Since the deposition of the calcium hydroxyapatite is layered in a manner similar to rings on a tree, it is possible to analyze past historical exposures. By examining the chemical composition throughout the layers, one can determine the relative timescale of exposure, including nutrient quality and chemical exposures. This line of work is extraordinary and demonstrates how technology can reveal new information that had previously been unmeasurable. This work complements other efforts to estimate the windows of susceptibility from childhood exposures. There are critical windows of development where nurture can have more influential effects (Wright, 2017). There are stages of rapid growth or development where one's exposome may have more detrimental or positive effects, and reconstruction of these exposures over time would be of great scientific benefit. This can be challenging, as biological samples are rarely collected regularly throughout development. More recent efforts have attempted to do just that and they are described in detail in Chapter 10, The Exposome in the Future.

3.12 Obstacles and opportunities

As described in Chapter 2, Genes, Genomes, Genomics: Advances and Limitations, our nature —that is, genetics— has taken center stage for more than a century. The driving forces that apply pressure to our genome are known to be important, but there has not been a similar level of effort into measuring these forces, especially within the biomedical science domain. This stands as one of the more challenging obstacles for exposome research. If we can get funders to accept that nurture is as important as nature, that the environment is critical to our health and survival, and that exposome science is creating tools to quantify this, we may be able to get more resources to study aspects of nurture.

In Darwin's time, breeders of plants and animals were quite sophisticated. New variations of orchids and crop plants were routinely generated through genetic crossings. Farm and domestic animals were carefully bred to create the ideal offspring. Darwin viewed this as artificial selection. He used the term natural selection as a distinction from artificial selection. Humans mated plants or animals in a manner that they saw fit, and Darwin viewed this as artificial or man-made. The use of the term natural selection came from what happens in nature prior to and without the input of human manipulation. It is unclear if we live in a state of natural selection as envisioned by Darwin. It would appear that we operate somewhere in between natural and artificial selection. Modernity has created many artificial pressures on health and reproduction. Regardless, the concept of selection is about how our environment and surroundings change and how we are able to respond and adapt to those changes. Nurture is what shapes us as a species and the exposome is attempting to measure as much of that as possible. Later chapters explore the techniques and approaches needed to capture this information.

We should avoid adversarial comparisons between nature and nurture. It is clear that the interaction between the two is what matters. Our analysis of nature has been bolstered by the advances in genomics described in Chapter 2, Genes, Genomes, Genomics: Advances, and Limitations, but the quantification of nurture has been elusive. The subtleties of nurture are immense, but many aspects of nurture can be captured by assessing what we have been exposed or subjected to, as well as examining how our bodies respond over time to the inputs of our environment. In this way a quantifiable exposome should be able to provide a numerical value of a major component of nurture, which should enhance our understanding of the nongenetic influences on health and disease.

3.13 Discussion questions

1. Consider your own education about nurture. Was the subject taught with the same detail and emphasis as nature (genetics)? If not, why do you think that was?
2. If you reflect upon your current state of health what do you see as the biggest drivers? Is it inherited traits or risk factors? Is it your childhood and adolescent upbringing? Is it your current behaviors and habits?
3. What are the factors that have been most critical to your nurturing? Were they positive or negative?

References

Barboza Solis, C., Kelly-Irving, M., Fantin, R., Darnaudery, M., Torrisani, J., Lang, T., & Delpierre, C. (2015). Adverse childhood experiences and physiological wear-and-tear in midlife: Findings from the 1958 British birth cohort. *Proceedings of the National Academy of Sciences of the United States of America, 112*(7), E738–E746. Available from https://doi.org/10.1073/pnas.1417325112.

Cecil, C. A., Smith, R. G., Walton, E., Mill, J., McCrory, E. J., & Viding, E. (2016). Epigenetic signatures of childhood abuse and neglect: Implications for psychiatric vulnerability. *Journal of Psychiatric Research, 83*, 184–194. Available from https://doi.org/10.1016/j.jpsychires.2016.09.010.

Chandler, J. D., Hu, X., Ko, E. J., Park, S., Fernandes, J., Lee, Y. T., ... Go, Y. M. (2019). Low-dose cadmium potentiates lung inflammatory response to 2009 pandemic H1N1 influenza virus in mice. *Environment International, 127*, 720–729. Available from https://doi.org/10.1016/j.envint.2019.03.054.

Chung, M. K., Kannan, K., Louis, G. M., & Patel, C. J. (2018). Toward capturing the exposome: Exposure biomarker variability and coexposure patterns in the shared environment. *Environmental Science & Technology, 52*(15), 8801–8810. Available from https://doi.org/10.1021/acs.est.8b01467.

Dennis, K. K., Auerbach, S. S., Balshaw, D. M., Cui, Y., Fallin, M. D., Smith, M. T., ... Miller, G. W. (2016). The importance of the biological impact of exposure to the concept of the exposome. *Environmental Health Perspectives, 124*(10), 1504–1510. Available from https://doi.org/10.1289/EHP140.

Epel, E. S., Blackburn, E. H., Lin, J., Dhabhar, F. S., Adler, N. E., Morrow, J. D., & Cawthon, R. M. (2004). Accelerated telomere shortening in response to life stress. *Proceedings of the National Academy of Sciences of the United States of America, 101*(49), 17312–17315. Available from https://doi.org/10.1073/pnas.0407162101.

Gage, F. H. (2019). Adult neurogenesis in mammals. *Science, 364*(6443), 827–828. Available from https://doi.org/10.1126/science.aav6885.

Gebreyes, W. A., Dupouy-Camet, J., Newport, M. J., Oliveira, C. J., Schlesinger, L. S., Saif, Y. M., ... King, L. J. (2014). The global one health paradigm: Challenges and opportunities for tackling infectious diseases at the human, animal, and environment interface in low-resource settings. *PLoS Neglected Tropical Diseases, 8*(11), e3257. Available from https://doi.org/10.1371/journal.pntd.0003257.

Goldstein, D. S. (2006). *Adrenaline and the inner world: An introduction to scientific integrative medicine*. Baltimore, MD: Johns Hopkins University Press.

Holzenberger, M., Dupont, J., Ducos, B., Leneuve, P., Geloen, A., Even, P. C., ... Le Bouc, Y. (2003). IGF-1 receptor regulates lifespan and resistance to oxidative stress in mice. *Nature, 421*(6919), 182–187. Available from https://doi.org/10.1038/nature01298.

Horvath, S. (2013). DNA methylation age of human tissues and cell types. *Genome Biology, 14*(10), R115. Available from https://doi.org/10.1186/gb-2013-14-10-r115.

Karlamangla, A. S., Singer, B. H., & Seeman, T. E. (2006). Reduction in allostatic load in older adults is associated with lower all-cause mortality risk: MacArthur studies of successful aging. *Psychosomatic Medicine, 68*(3), 500–507. Available from https://doi.org/10.1097/01.psy.0000221270.93985.82.

Kenyon, C., Chang, J., Gensch, E., Rudner, A., & Tabtiang, R. (1993). A *C. elegans* mutant that lives twice as long as wild type. *Nature, 366*(6454), 461–464. Available from https://doi.org/10.1038/366461a0.

Kuhn, H. G., Toda, T., & Gage, F. H. (2018). Adult hippocampal neurogenesis: A coming-of-age story. *The Journal of Neuroscience, 38*(49), 10401–10410. Available from https://doi.org/10.1523/JNEUROSCI.2144-18.2018.

Lee, S. J., Murphy, C. T., & Kenyon, C. (2009). Glucose shortens the life span of *C. elegans* by downregulating DAF-16/FOXO activity and aquaporin gene expression. *Cell Metabolism, 10*(5), 379–391. Available from https://doi.org/10.1016/j.cmet.2009.10.003.

Masoro, E. J. (2005). Overview of caloric restriction and ageing. *Mechanisms of Ageing and Development, 126*(9), 913–922. Available from https://doi.org/10.1016/j.mad.2005.03.012.

McEwen, B. S. (1998). Stress, adaptation, and disease. Allostasis and allostatic load. *Annals of the New York Academy of Sciences, 840*, 33–44. Available from https://doi.org/10.1111/j.1749-6632.1998.tb09546.x.

Morishita, H., & Arora, M. (2017). Tooth-matrix biomarkers to reconstruct critical periods of brain plasticity. *Trends in Neurosciences*, *40*(1), 1–3. Available from https://doi.org/10.1016/j.tins.2016.11.003.

Murphy, C. T., McCarroll, S. A., Bargmann, C. I., Fraser, A., Kamath, R. S., Ahringer, J., ... Kenyon, C. (2003). Genes that act downstream of DAF-16 to influence the lifespan of *Caenorhabditis elegans*. *Nature*, *424*(6946), 277–283. Available from https://doi.org/10.1038/nature01789.

Seeman, T. E., McEwen, B. S., Rowe, J. W., & Singer, B. H. (2001). Allostatic load as a marker of cumulative biological risk: MacArthur studies of successful aging. *Proceedings of the National Academy of Sciences of the United States of America*, *98*(8), 4770–4775. Available from https://doi.org/10.1073/pnas.081072698.

Watts, D. J. (2003). *Six degrees: The science of a connected age* (1st ed.). New York: Norton.

Wright, R. O. (2017). Environment, susceptibility windows, development, and child health. *Current Opinion in Pediatrics*, *29*(2), 211–217. Available from https://doi.org/10.1097/MOP.0000000000000465.

CHAPTER FOUR

The environment: the good, the bad, and the ugly

The more clearly we can focus our attention on the wonders and realities of the universe, the less taste we shall have for destruction.

Rachel Carson, The Silent Spring

4.1 Welcome home exposome

The preceding chapter provided an overview of nurture from a broad perspective. Many key components of nurture, especially within the context of the exposome, fall under our current model of environmental health sciences. This chapter will serve as an overview of the environment with a specific focus on how humans interact with their surroundings and will look at how these topics are currently studied. Whereas Chapter 3, Nurturing Science, examined these broad external pressures over human history, this chapter will focus on modern environmental issues from more of a post-Industrial Revolution perspective. It will address the deleterious effects of poor diets, lack of activity, or other detrimental exposures from environmental sources, such as persistent organic pollutants, heavy metals, and particulate matter. It will also highlight potentially positive aspects of the environment, such as good nutrition, the built environment, and helpful social interactions.

Although the environmental health sciences field has not fully adopted the expansive view of the exposome, the principles of both areas are closely aligned, and I expect we will see more exposome-based courses and curricula in the coming years. Based on decades of technical and conceptual expertise in analyzing environmental influences on health, it is reasonable to consider that the primary intellectual home of the exposome will be environmental health sciences. With the focus on how our environment influences health and disease, exposome-based approaches have the potential to greatly improve our understanding of these effects. In

The Exposome.
DOI: https://doi.org/10.1016/B978-0-12-814079-6.00004-3

general, environmental health sciences encompasses research conducted at the population level via environmental epidemiology, assessment of actual environmental exposures via exposure science, and examination of the biological alterations induced by environmental factors via toxicology. These three areas, which span populations to molecules, comprise environmental health sciences (Fig. 4.1) and provide a framework for exposome research. There are other components of environmental health sciences, such as risk assessment and disease ecology, that complement the exposome paradigm. There have been notable enhancements in the application of data science and other computational tools to evaluate complex exposures, and these are explored in Chapter 8, Data Science and the Exposome.

The field of environmental health sciences is at its best when it capitalizes on the transfer of knowledge among the three major divisions within the field. Toxicologists and exposure scientists rely on epidemiological studies and analysis of exposures to determine which compounds should be studied. Epidemiologists rely on toxicologists to determine if their observed associations are in line with what is known about the biological and toxicological pathways involved in the health effects, and on exposure scientists to document the actual exposures. Toxicologists also look to the exposure scientist to know what the relevant exposure levels are. When data from all three subdisciplines converge, the results provide a very strong scientific basis for that particular environmental exposure having an impact on human health, including identifying causal relationships. A

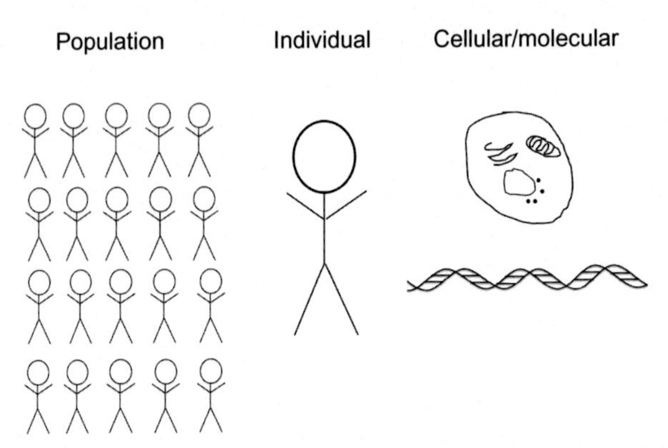

Figure 4.1 Populations to people, cells, and molecules. Exposome-related data will come from a broad range of sources, including population studies, personal environmental and biological monitoring, and laboratory studies of animals, cells, and molecules.

similar level of convergence among all components of environmental health sciences will be necessary for data from the exposome to have an impact on human health (Fig. 4.2).

One of the challenges for environmental health sciences is that, historically, environmental exposure studies have been focused on those exposures that are unintentional. Intentional ingestion of drugs or alcohol, cigarette smoking, overeating, and levels of physical activity typically fall outside the domain of the field. Separating exposures that have a volitional component from those that are passive is a reasonable structure when determining allocation of research resources and addressing sources of exposure. However, when examining the health of the individual, it becomes difficult to separate out these exposures because they are so interrelated. Exposome research resists these artificial distinctions and strives to integrate as much information as possible. Historically, environmental health and occupational health have worked together. Occupational health training programs are often part of environmental health departments and many have both of the words "occupational" and

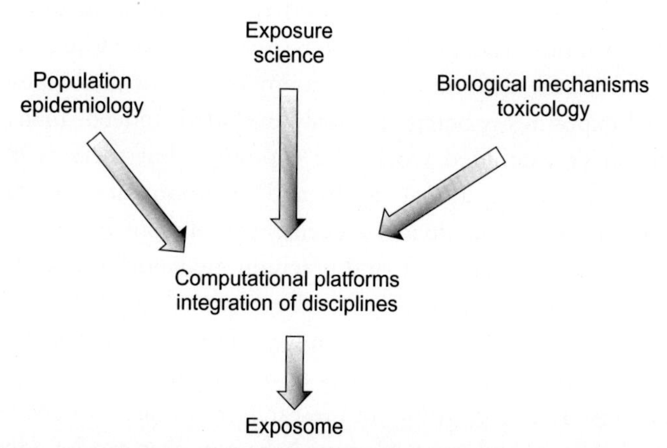

Figure 4.2 The components of environmental health sciences and the exposome. Each of the subdisciplines of environmental health sciences can contribute to the elucidation of the exposome. Information and findings from populations (via environmental epidemiology), accurate measurement of chemical species in our surrounding media (exposure science), and data on biological mechanisms and target pathways (via toxicology and biochemistry) will all be needed for exposome research. The findings from the three major subdisciplines need to be integrated via computational methods, along with data from atmospheric chemistry, genomics, behavioral science, nutritional sciences, and many others.

"environmental" in their title. This has occurred at the departmental structure and training programs, and while many programs have removed the word "occupational" from their title, they still include the work under their structure. Thus, exposures (again, primarily unintentional) that occur in the home, community, and workplace continue to fit under environmental health sciences in a logical form.

Within the U.S. NIH, the study of drugs of abuse, such as cocaine, amphetamines, nicotine, and opiates, fall under the National Institute of Drug Abuse, while alcohol consumption falls under the National Institute of Alcohol Abuse and Alcoholism. Thus, research on these major environmental exposures are removed from the mission of NIEHS, which oversees research in environmental health sciences. In a similar manner, workplace exposures generally fall under the National Institute for Occupational Safety and Health (NIOSH). Although NIOSH does fund some research, it is not part of NIH and is separated from the major biomedical research mission. Exclusion of exposures from the occupational setting within the biomedical enterprise is somewhat troubling for the individual. This is not to say that exposures that occur in the workplace are off limits for research in environmental health sciences, but rather the organizational structure is not optimized to include occupational exposures and prevention strategies into the research agenda. When occupation is removed from the definition of environment, one of the biggest sources of potential exposures is being arbitrarily excluded. Indeed, many important studies have examined workplace exposures, but there is still a disconnect between occupational health and environmental health from a research perspective. This divide is even more obvious between occupational and environmental health and precision medicine and healthcare.

Obesity is addressed primarily by the National Institute of Diabetes and Digestive and Kidney Diseases, and exercise and physical activity do not have a specific home. As the rates of obesity grow, it is disappointing to not see more resources going into research in this area. Increased physical activity is one of the most effective behaviors that can be adopted for better health. From a biomedical perspective, we do know enough about the effects of exercise to encourage it at a population level. For many chronic medical conditions – for example, in cardiac rehabilitation and metabolic disease – we also have a substantial amount of knowledge. Yet, we do not know enough about what motivates people to exercise. It is more than simply going to the gym. As discussed in the previous chapter, there are several modalities to increase physical activity. This is where the built environment and city planning/urban design can play important roles, along with school-based programs, to encourage physical activity. When systems are designed to increase physical activity, the effects are seen at a population scale. However, these topics do not fall under the domain of the National Institutes of Health. As many have noted before, NIH might more appropriately be named the National Institutes for Disease. NIH studies diseases and develops strategies to treat the diseases and to prevent them to some degree. NIH has had some remarkable success in this area and I do not mean to be dismissive. However, when you are trying to affect the health of over 300 million people, a more proactive and preventative approach is necessary. Similarly, when we look at global health, we must adopt preventive strategies that can impact billions.

Regardless, a focus on unintentional environmental exposures is the charge of the field of environmental health sciences, but the field must be receptive to other exposures that may be due to voluntary activities or occupations. Another way to think of it is that the field of environmental health sciences should focus on the *measurement* of classical environmental exposures, but be *inclusive* when it comes to other types of data when interpreting the former. By inclusive I mean the deliberate inclusion of these types of data in the studies and analyses. This is because our environmental exposures occur within the context of these other dietary and lifestyle exposures, and ignoring these coincident exposures could seriously confound results (intellectually, not statistically; the converse is a real problem that I discuss in Chapter 8, Data Science and the Exposome). For example, a community that resides near a factory may have a higher incidence of a particular disease and if one only focuses on those potential exposures from the factory and does not take into consideration the various socioeconomic, dietary, and lifestyle factors, important drivers of the disease could be missed. An elevation in the lipid-soluble chemical could well occur, but the reason for the increased disease risk could be because of the consumption of a high-fat diet or a particular cuisine and not necessarily the concomitant increase in the chemical. The chemical may just be a biomarker of an unhealthy diet (if everyone that lives near the factory has poor access to food and bad dietary habits). Of course, environmental epidemiologists control for most of the socioeconomic and lifestyle factors, but it is extremely difficult to control for complex biochemical factors that include variations in cytochrome p450, redox pathways, and fluctuations in hormonal systems. For the most part, these factors are controlled for using questionnaire and census-type data, not biological data.

Having parallel laboratory analysis to assess nutritional and redox status, confirm smoking status, and measure genetic vulnerability would be beneficial in determining the appropriateness of the control group. Further, with the availability of improved measures of exposure and biological response, the traditional survey instruments can be refined and strengthened to improve their predictive value. Ideally, we will not only focus our attention on chemicals that have an adverse health effect and work to remediate them, but also address other components such as poor nutrition and low physical activity.

In the past few decades, scientists have studied chemicals suspected of causing human disease and have been able to take steps to reduce exposures. This has required measuring these chemicals both inside the human body and in the general environment. These techniques are explored more in the next chapter, but it is important to note that we do not know which specific chemicals to focus on. Tens of thousands of chemicals are in production throughout the world, and it is simply impossible to rigorously test them one by one. Moreover, since we are exposed to them in combination, there is an experimental reality that makes testing of these combinations seemingly impossible. This is why the newer non- or untargeted approaches hold so much promise. The idea is to test every possible chemical signal that is detectable, even if the identity is currently unknown. Exposure scientists have worked hard to optimize sensitive and highly accurate methods for known toxicants. Toxicologists work with rigorously tested compounds with minimal impurities. Unfortunately, this is not the situation we face in our day-to-day exposures. There are more unknowns than knowns, and we need a way of examining the unknowns in a systematic fashion.

Environmental health scientists must be the drivers of the exposome—it is essential that those driving the work have the appropriate expertise in the environmental issues. Ironically, investigators from other fields have often been more receptive to this concept than many of those within the field. Reluctance in adopting an approach that is counter to much of one's training is not surprising, but the receptiveness of those outside the environmental health community indicates that there is something about the exposome that resonates with the broader biomedical and scientific enterprise in a way that traditional environmental health sciences does not. Although I view the exposome as much more than a mere repackaging of environmental health, it may be that the field could benefit from some new branding and packaging.

It is not to say that we abandon our traditional approaches; in fact, we likely need to conduct more of the traditional type of research. But the exposome demands that we draw data from outside our historically defined intellectual boundaries. We do not have to necessarily measure all of these other aspects; however, we need to use data that are already available and include them in the interpretation of our studies. As it turns out, many of our analytical approaches can and do measure the variables that lie outside of environmental health, so why not include them in our analyses?

4.2 Toxicology—mechanisms of toxicity

The concept of the exposome was essentially laid at the feet of exposure assessors and epidemiologists. The argument was that epidemiologists needed better and more comprehensive data from the exposure assessors. It was up to the field of exposure assessment/science to deliver the goods. There was no obvious place for the toxicologist. If one accepts the more expansive definition of the exposome–that is, the exposome is representative of all external forces, not just exposures per se, but also the body's responses–the role of the toxicologist becomes readily apparent. In fact, the mechanistic understanding of environmental agents on biological systems becomes foundational.

Toxicology is defined as the study of the adverse actions of chemicals on biological systems. This is a very broad definition that includes every type of chemical and every type of biological system, from plants to bacteria, to fish, and to humans. Toxicology aims to understand the biological consequences of exposures, from environmental chemicals, pharmaceuticals, and other contaminants. For many years, toxicology was a descriptive science that merely cataloged how toxic each entity was. For example, toxicologists would determine the dose of a particular chemical that kills half of the study population—the rat LD_{50}, the dose at which 50 percent of a given laboratory animal group was killed by the chemical. This is useful information, as it allows the comparison of thousands of chemicals for a measure of toxicity. Over time, toxicology went from death as an endpoint to more subtle, disease-related endpoints and the identification of the molecular processes that explained the toxicity. For the purpose of this book, I focus the discussion of toxicology on the adverse effects of chemicals on human biology. The importance of toxicology to exposome research is that it provides information on the biological effects of toxicants. Toxicologists determine how chemicals interact with organ systems, specific pathways, cells, and macromolecules. I have received training in neuroscience, pharmacology, physiology, and biology, but I view myself as a toxicologist and admit to having a bias in support of the field.

Based on my experience, I believe that toxicology should adopt the exposome as a means of integrating its mechanistic data into models of human health. This is not to say that toxicologists need to dramatically change what they are doing, but rather they need to position themselves to be able to contribute their findings to the exposome paradigm. The biological pathways affected by the toxic compounds are a key component of exposome-based research. I recall looking at epidemiological data as a student to see what exposures had been associated with a disease. This is a tried and true approach, but it puts great dependence on how that epidemiological study was conducted. Since an expansive, exposome-based approach was not taken at that time, this strategy was anchored in the proverbial "looking under the streetlamp" approach. If an epidemiological study examined 15 different endocrine disruptors, that study would reveal data on only those 15 compounds. A key metabolite or downstream modifier could go unidentified. In a similar vein, Dr. Thomas Hartung at Johns Hopkins University has suggested that a human toxome should serve as a framework for study within the field of toxicology (the creation of another -ome notwithstanding, the adoption of an unbiased approach is welcome). The Human Toxome (human-toxome.com) is analogous to the exposome, but with a specific focus on the adverse effects of toxic environmental chemicals. One could view the toxome as a specific subset of the exposome. The findings from such an effort should easily be incorporated into the exposome framework. (Note: this academic project should not be confused with the Human Toxome Project being conducted by the Environmental Working Group (EWG) and Commonweal in California, which is focused on measuring specific chemicals in a relatively small number of people. The apparent goal of the EWG initiative is to raise awareness about the presence of these chemicals, but due to its limited experimental design, it is unlikely that the data generated will be of value to exposome efforts or the scientific community at large—as it is not -omic scale research). One could argue that Hartung's narrower approach is more in line with the traditional view of environmental chemical exposures. The types of data collected by this initiative, especially the identification of pathways of toxicity (PoTs), are laying the foundation for a major step forward for toxicological risk assessment, but they could also be essential for the interpretation of exposome data. The toxome approach and the Tox21 initiative (discussed later in this chapter) should be encouraged in that they are generating data that have the potential to add to the body of the exposome. It will be important to pursue these studies with the mindset that the findings can be absorbed into the exposome framework. For example, if a toxome approach identified 25 different pathways involved in carcinogenesis, it should be possible to build a computational model that allows the screening of combinations of chemicals for their ability to disrupt these pathways and contribute to regulatory risk assessment. The computational nodes could become integrated into a larger exposome model, and data from exposome efforts could be used to identify the combinations of chemicals to be tested in the PoTs identified in the toxome efforts.

Another critical aspect of toxicology is that it provides information on the biological mechanisms of the adverse exposures. From the use of curare to understand the neuromuscular junction, to the work of the late Toshio Narahashi using tetrodotoxin to understand sodium channel function, toxins have been used throughout history to uncover biological secrets (Narahashi, Anderson, & Moore, 1966). Narahashi went on to use insights from the naturally occurring tetrodotoxin to determine the actions of the pesticide DDT and many other synthetic chemicals (Narahashi & Haas, 1967). Learning about biology by paying attention to the lessons taught by nature is a recurring theme throughout science. Understanding how the various environmental toxicants alter human biology is critical to understanding disease pathogenesis and can identify molecular targets and pathways that can be targeted for intervention.

The phrase "the dose makes the poison" is a useful construct for explaining the concept of dose–response. Distilled from the words of Paracelsus, the Father of Toxicology, the phrase elegantly reveals the relationship of the dose of a compound and its effects (adverse in this case). As I traveled through Austria and Hungary, I came across several exhibits and stories of his work—from his training at the University of Vienna to his burning of Galen's texts in front of the university (a very Lutheran act, which given the century was not unexpected). Unfortunately, the toxicological catchphrase "the dose makes the poison" is unable to capture the nuances and shape of the dose–response curve. Decades ago the relationship was thought to be not far from linear. For the most part the phrase is true, but there are subtleties that should not be overlooked. Paracelsus taught us that good things can be bad at high doses, so the thought that bad things can be good at low doses seems counterintuitive. But in fact, there is strong evidence to support the concept of hormesis for many classes of compounds and some biological pathways. Hormesis is the term used to describe the phenomenon where low doses of a compound that is demonstrably toxic at higher doses may actually provide a beneficial effect. Low levels of oxidative stress activate complex response pathways that make it easier for the body to respond to subsequent insults. While it is a bit speculative, this could be part of the reason that antioxidants routinely fail in clinical studies of disorders that are thought to involve oxidative stress in their pathogenic process. Low levels of some stressors may be beneficial to the organism. Ed Calabrese has worked tirelessly over the past decade or so to bring the concept of hormesis back into the dose–response discussion (Agathokleous & Calabrese, 2019). A basic scientist can consider hormesis and look for its qualities in laboratory studies. However, the concept of hormesis can cause apoplexy in the regulatory toxicologist. When making decisions about safe levels of environmental chemicals, the idea that there may be a beneficial effect of obviously toxic compounds (U-shaped curves) just does not fit into the mental construct of regulatory science. Dr. Calabrese and others have documented the hormetic effects in many situations. It appears that radiation may even have some hormetic effects. This is very challenging to the dogma that any exposure to a mutagen is bad. So what happens when someone is exposed to dozens or hundreds of chemicals theorized to have a hormetic effect? If the chemicals are related, do the effects become additive pushing the curve past the positive to the adverse? This is where approaches that look at the combination of chemicals to which we are exposed could provide needed insight. The exposome is open to this concept in that much of the work will proceed in an unbiased and hypothesis-free format.

Computational toxicology is a more recent development within the field of toxicology. While most are familiar with the quantitative structure-activity relationships used in drug development, a similar approach can be used in toxicology to model potentially toxic interactions. The U.S. Environmental Protection Agency (EPA) has established a Computational Toxicology Center to develop and apply computational tools to aid with the prediction of toxicity as it relates to regulatory decision-making. While I was writing the first edition of this book, these efforts were in a relatively early stage, but now we are seeing the full force of the computational advances be applied to the study of environmental chemicals. I expanded Chapter 7, Pathways and Networks, and Chapter 8, Data Science and the Exposome to capture these exciting approaches. These advances are fortuitous for exposome research. The fact that only a small fraction of chemicals introduced into our environment have undergone toxicological testing highlights the need for faster and less expensive means of assessing toxicity. The goal is to develop predictive toxicity models that allow for the identification of the compounds with the highest likelihood of being toxic so that they can undergo more extensive testing. Several major initiatives are yielding exciting results and will ultimately be major contributors to the exposome even if they do not yet realize it. When coupled with recent advances in high-throughput technologies, the potential for expanding the number of compounds screened or tested is very high.

The idea of high-throughput toxicology is that we need to gain information on thousands of chemicals, but most classical toxicological studies examine compounds one by one. It is exceedingly difficult and expensive to test compounds in the types of studies required by regulatory agencies. If classes of chemicals could be tested under the same conditions, it may be possible to extrapolate potential adverse effects. Many assays on cellular growth or cellular injury can be performed in 96, 384, or 1436 well trays using robotics. This allows for the high-throughput-type approaches that are being conducted under Tox21, a joint venture among NIEHS, EPA, the National Center for Toxicology Research at the Food and Drug Administration, and the National Center for Chemical Genomics at NIH. Tox21 set an ambitious goal to test over 10,000 chemicals on a battery of high-throughput assays, providing data on compounds that have never undergone regulatory analysis. Such data can be used to develop and validate computational models that may permit in silico testing of hundreds of thousands of compounds. At this time, they have achieved their goal by testing over 10,000 chemicals in more than 70 biological assays (https://www.niehs.nih.gov/health/materials/tox21_508.pdf). This is an extraordinary dataset that is a gift to the exposome.

Toxicologists who conduct research in the biomedical (vs. the regulatory) side are being challenged by funding agencies to demonstrate the translational nature of their research. For many the connection is not obvious; however, the exposome provides a useful translational conduit. If findings from toxicology studies make their way into the exposome paradigm, which they should, then by its very nature, the research will be translational as it is being incorporated into a model aimed at improving human health. These newer approaches have extraordinary potential, and the pathway and network models are described in more detail in Chapter 7, Pathways and Networks.

4.3 Exposure science (or assessment, or biology)

For decades the field of exposure assessment has been determining the composition of our surroundings and how various contaminants enter the body. More recently, the term "exposure biology" has been used to describe the study of the biological aspects of exposures and is one that biomedical funding agencies seem to prefer in that it is more aligned with their biologically based approaches. From the standpoint of environmental health sciences and the exposome, the more expansive and inclusive "exposure science" would appear to be the best descriptor of dealing with exposures. Exposure science not only encompasses the measurement of contaminants in our air and water, but also includes measuring the chemicals within the body or other biomarkers of exposure. For example, the chemical composition of an airborne pollutant and its physicochemical interactions with other airborne chemicals would not necessarily be part of exposure biology but would clearly be of importance and fall within the domain of exposure science. This is similar for exposure assessment. Assessing what is in the environment is only one component of exposure science. Without an understanding of how chemicals move from the environmental matrix to the target species, it is difficult to predict the biological consequences.

The field of exposure science/assessment was one of the first to grapple with the exposome concept. Indeed, Wild threw down the proverbial gauntlet at the feet of the field of exposure science. Paul Lioy, a leader in the field of exposure assessment, took umbrage at Wild's request. He argued that exposure science was already generating the appropriate information and suggested that semantic arguments about terminology were misguided, and that the field was wasting time by not doing the actual science (Lioy & Rappaport, 2011). While I agree that conducting the science is the ultimate goal, the framing of the questions and scope of the problem, which includes definitions, is not trivial. However, I would argue that if it is framed correctly, the exposome has the potential to attract more attention and financial support for the evaluation of the environment in our health. At the same time, Steve Rappaport and Martyn Smith from the University of California at Berkeley were actively embracing the exposome concept. Working from their NIEHS-funded Superfund Project they proposed that the exposome could advance and provide critical direction for the field (Rappaport & Smith, 2010). Rappaport has also proposed that the measurement of DNA and protein adducts could represent biological measures of cumulative chemical exposures. Rappaport and Smith have been key proponents and leaders of the exposome concept. From the perspective of an exposure scientist, such as Lioy, it is understandable why a focus on internal biological markers of exposures would be irritating. If one believed that the exposome would only rely on internal biological markers, there would be a reason for concern. But this is not the case. The exposome, as defined here and by Wild, incorporates the rich exposure data provided by the field of exposure science, as well as the data collected from human samples. In fact, the environmental exposure data are critical in the interpretation of the internal biological markers of exposure and response.

Lioy had suggested the use of the term "eco-exposome" to reinforce the importance of our surrounding environment, including the health of the planet. This concept was also introduced in the National Academy of Sciences report on Exposure Science for the 21st Century, which Lioy cochaired. The working group proposed the use of the term "eco-exposome" to explain the exposures that reside outside of the body, but the revised definition of Miller and Jones cited in Chapter 1, The Exposome: Purpose, Definitions, and Scope, is all-encompassing, making the need for such a distinction obsolete. The point of the exposome is to act as an interface between the external environment and human biology. I had the pleasure of having a spirited informal debate on this topic with Lioy during a systems biology workshop at Georgia Tech. I obviously argued against the introduction of an exposome derivative. Unfortunately, Dr. Lioy died shortly after this meeting, preventing further discussions with him on this topic. His death has been a notable loss for the scientific community.

As I outlined in the first chapter, creating additional permutations to a concept that is already battling intellectual inertia will just weaken the case. Investigators may focus on one specific aspect or subset of the exposome, such as those described as part of the eco-exposome, similar to how a person in the field of proteomics will only study one specific posttranslational modification. However, there is not a posttranslational proteome, just a proteome. Similarly, it would be preferable to incorporate all aspects into a unified definition of the exposome, lest we provide more fodder for the -omic critics. Bert Brunekreef at Utrecht University has also addressed these distinctions, noting that the definition of the exposome put forth by Rappaport and Smith is narrower versus the more expansive and inclusive view of Wild (Brunekreef, 2013). Brunekreef also noted that the eco-exposome broadens the exposome sphere even more. I agree that the health of the planet is a key component to our own health, but ecological concerns, such as weather patterns, climate change, and plant health are outside of the working model of the exposome, which is focused on human health. These are factors that influence health and exposures, but under the concept of balancing our understanding of human phenotypes ($G \times E = P$), exposome research should be anchored in human health and not try attempt to study planetary or ecological issues.

To understand the importance of measuring environmental contaminants outside and inside the body, consider the soup analogy (Fig. 4.3). Imagine tasting the following items individually: raw carrots, raw chicken (remember you are just imagining this, no need to worry about salmonella), raw potatoes, raw onions, salt, and pepper. Now describe the taste of the soup that will be made from these ingredients. It is difficult, especially if you have never had chicken soup. What if the amount of onions were doubled, or the pepper halved? Until you taste the final compilation that has simmered for several hours, you will not know what it will taste like. Assessing the impact of the environment on our health is similar. Measuring the individual components outside of the body cannot tell you what the soup will taste like, no more than tasting the soup can allow you to describe the precise ingredients and quantities used to make the soup itself. In order to truly understand the soup, you must know what goes into the soup *and* what the end result tastes like. The exposome is useless without the combined knowledge of the individual ingredients that go into it and an assessment of the merged ingredients within the biological system. If it is determined that a particular combination of exposures increases the incidence of the disease, knowing the components in our environment can inform the steps to mitigation. To take this

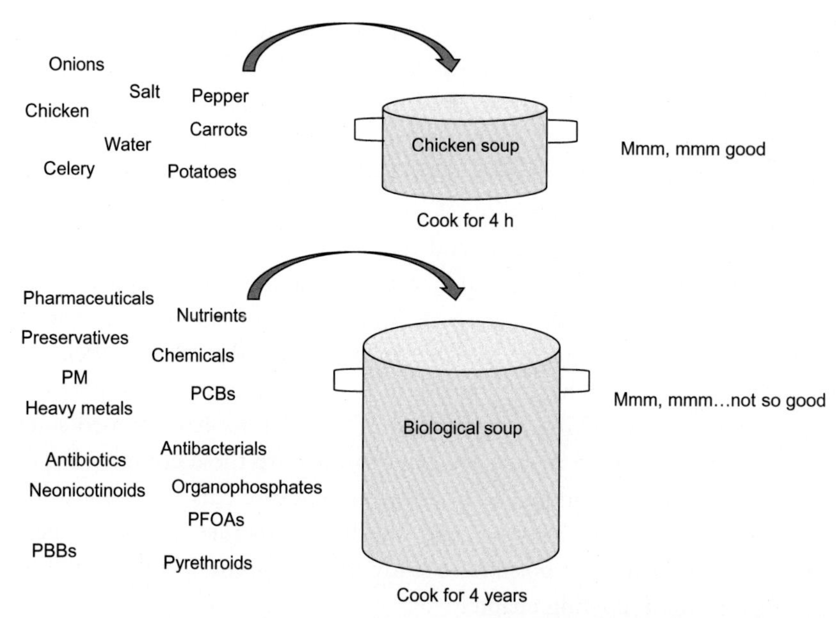

Figure 4.3 The soup analogy. To make soup, individual ingredients are collected, prepared, and added into a pot with water and simmered for hours. One can view our bodies (and bloodstream) as a type of soup composed of the various ingredients we add through our exposures. If one samples the individual ingredients of the soup, it may be possible to estimate what the end result will be, but it will be impossible to know exactly how it will taste. Until the ingredients are put in the pot together and are subjected to the cooking process, which involves metabolic conversion and molecular interactions of the ingredients, the ultimate flavor of the soup is not apparent. Likewise, sampling the individual ingredients outside of our bodies cannot tell us what the essence of our exposures is. Sampling the final product will allow us to estimate what ingredients went into it, but not necessarily their forms and amount. However, if one conducts a thorough analysis of the individual ingredients before they are added and then conducts a similarly detailed analysis of the end result, it is possible to determine the level of transformation and interaction of the various ingredients after entering the body. A major goal of the exposome is learning how the various ingredients and processes impact the flavor of the soup so that we can make the soup taste better. *PBB*, Polybrominated biphenyls; *PCB*, polychlorinated biphenyls; *PFOA*, perfluorooctanoic acid; *PM*, particulate matter.

cooking analogy a bit further, imagine if a group of 10 people were given the recipe for this soup. Would the end result be the same for all 10? Unlikely. Some would cook it longer, dice the carrots up finer, cut the chicken into different shapes, or use more or less salt. The individual variability in cooking style pales in comparison to the individual variability of our biology. Just because several individuals are exposed to the same ingredients does not mean that they will end up with the same soup. In fact, it is highly unlikely that they will (Fig. 4.3). Therefore, you must conduct a quantitative analysis of the soup and the ingredients.

The implementation of more person-specific exposures and responses, the types of approaches being discussed for the exposome, are necessary to address the individual variability. When we have information from personal environmental monitoring and biological measures, which incorporates individual variability, and then combine those data with environmental monitoring station data, a powerful set of information begins to emerge. It becomes possible to determine the relative effects of individual activities, behaviors, and housing on exposures. My previous resistance to the launching of new field of exposomics has softened since we are now developing new approaches to measure these complex exposures that include individual and population exposure data in combination with biological modifications and health outcomes. Indeed, new approaches are making it possible to study the exposome and these will be described in the following chapter.

4.4 Epidemiology and the exposome

I suspect that the environmental epidemiologists will be the last group to embrace the exposome. The epidemiologist needs validated measures of exposures. Unidentified mass spectral peaks are not any better than the real measures that exist today. In fact, I would be concerned if this group readily embraced the exposome without appropriate validation. It is up to the exposure scientists, toxicologists, and others in the biological arena to demonstrate that an exposome-type measure is truly measuring an outcome of biological importance. Once this is established, then it will be incumbent upon the epidemiologist to use the measures that provide more detailed information about exposures and responses.

Enhanced quantitative information on exposures will improve the ability of epidemiological studies to identify the underlying cause of disease. The exposome is ultimately an epidemiological construct. The textbook definition of epidemiology is "a branch of medical science that deals with the incidence, distribution, and control of disease within a population" or "the sum of the factors controlling the presence or absence of a disease or pathogen." Environmental epidemiology focuses on the environmental factors that contribute to the incidence, distribution, and control of disease within a population. Thus, the ascertainment of the environmental factors is essential. To date, there has not been a means of capturing the impact of complex mixtures, which is something the exposome can address. Epidemiology uses extremely well-validated approaches and rigorous frameworks, such as the Bradford-Hill criteria. As such, their conclusions are statistically strong, but often overly conservative. The associations must be strong enough to achieve significance, but like GWAS if there are chemicals/genes that have an effect in only 10 percent of the population, these genes will often go undetected. In order for exposome data to be integrated into epidemiological models, the data cannot exist as 100,000 unidentified peaks from high-resolution mass spectroscopy. As discussed in the later chapters, there will need to be some data complexity reduction. For example, an exposome risk score analogous to a polygenic risk score could be one possible approach.

A major focus of epidemiological research focuses on infectious agents. Such agents are amenable to studies of transmission and spread of disease. The nature of infectious agents is more tangible, in that the pathogens can be identified via polymerase chain reaction (PCR), sequencing, or enzyme-linked immunosorbent assay (ELISA), and their vectors can be measured and tracked. While some environmental factors are somewhat easy to assess, many others are much more difficult. For example, determining one's smoking history is fairly straightforward. However, determining one's exposure to a particular pesticide can be daunting. Individuals do not perform well when recalling their past exposures, and interviews and survey instruments only address the potential exposures, not the actual biological exposures. This is where the field of exposure science has had a major role and where Wild encouraged improvements. By directly measuring the chemicals that are in an individual's or a population's environment, it is possible to estimate the exposure. The measurement of what is the environmental matrix is highly accurate. However, translating the quantity of chemical in the environment to what ultimately ends up in a person's body is much less definite. Indeed, the individual variability in uptake, absorption, distribution, metabolism, and excretion is vast. Attempts to measure all of the constituents in a biological fluid such as plasma have the potential to measure internal dose, metabolism, and biological effect at the individual level, essentially capturing these individual variabilities all at once. By measuring the compilation of chemicals and metabolites in the human body and comparing that to the concentrations in the surrounding environment, one can start to address the issue of individual variability and determine what the most accurate predictors of chemical deposition are in an individual's bloodstream.

In the landmark Global Burden of Disease report, over a dozen risk factors are shown with their relative impact on disease (GBD, 2016, 2017). It must be noted that this is a global evaluation. If one focused on one specific country, there would be notable shifts in many of these factors. Historically, indoor air pollution and ambient particulate matter would have been addressed by environmental health scientists, but most of the others would be considered to be outside their domain. However, the exposome captures all of the risk factors addressed in the Global Burden of Disease project. Smoking, alcohol consumption, and low physical activity are part of the lifestyle or behavior component. High blood pressure, high body mass index, high fasting glucose, and high cholesterol can all be viewed as part of the cumulative biological response. Low fruit intake, low nut and seed intake, high sodium, and iron deficiency are all captured under dietary influences. All of these risk factors are part of the refined definition of the exposome. They are the modifiable components that could well be part of a comprehensive exposome index. This study is a prime example of how the exposome could be used to expand the understanding of disease on a global scale.

4.5 The chemicals

Synthetic chemicals are found in a wide variety of consumer products, building materials, industrial goods, pharmaceuticals, and foods. Many of these chemicals were designed and synthesized for a specific reason, whether it be to inhibit microbial growth, make containers more durable, kill harmful insects, extend the shelf-life of foods, slow growth of cancer cells, or prevent flammability. For decades the chemical industry's slogan was "better living through chemistry." It was useful marketing and scientifically accurate to a point. When chemicals are used in a specific context in a targeted way, they truly can improve our lives. Unfortunately, chemicals do not restrict themselves to their original intent and site of action. Diffusion rules. Chemicals leach out of plastic products, car exhaust spreads through neighborhoods and cities, pesticides drift after spraying, and drugs leave the system for which they were intended and find their way into water systems. Better living through chemistry may be reasonable with the caveat that the chemical does not indiscriminately enter our environment where it can be inhaled, ingested, or absorbed. Since we cannot control chemical diffusion and distribution, we must conclude that all chemicals that are used in commerce will enter the environment. It is not just the chemicals coming from smokestacks. It is all of them. There are hundreds of books, websites, and databases that catalog the breadth of chemicals that have been in commercial production, which we now consider to be part of our environment.

My exposome-centric thinking makes it difficult to make a list of chemicals of concern. The Stockholm Convention of Persistent Organic Pollutants has identified a list of chemicals deemed to be the most harmful to human health. I do not disagree with their list; it is just too short (even with expansion). It follows traditional toxicological knowledge and does not take into account mixture effects and the combined exposure to thousands of chemicals. At the same time, it does highlight some of the most persistent and dangerous chemicals. Many contain one of the halogens- chlorine, fluorine, or bromine which contribute to their persistence and toxicity.

The original 12 chemicals from the Stockholm Convention (pops.int)

Aldrin
Chlordane
DDT
Dieldrin
Endrin
Heptachlor
Hexachlorobenzene
Mirex
Toxaphene
Hexachlorobenzene
Polychlorinated biphenyls
PBDEs
Polychlorinated dibenzofurans

The next round of 16

Alpha hexachlorocyclohexane
Beta hexachlorocyclohexane
Chlordecone
Decabromodiphenyl ether (commercial mixture, c-DecaBDE) hexabromobiphenyl

Hexabromocyclododecane
Hexabromodiphenyl ether and heptabromodiphenyl ether (commercial octabromodiphenyl ether)
Hexachlorobutadiene
Lindane
Pentachlorobenzene
Pentachlorophenol and its salts and esters
Perfluorooctane sulfonic acid, its salts and perfluorooctane sulfonyl fluoride
Polychlorinated naphthalenes
Short-chain chlorinated paraffins
Technical endosulfan and its related isomers
Tetrabromodiphenyl ether and pentabromodiphenyl ether (commercial pentabromodiphenyl ether)

It is not difficult to agree that these heavily used, widely disseminated, and environmentally persistent chemicals are bad for human health and our environment. However, there are over 80,000 synthetic chemicals in production. Coming up with a top 10 or 20 does not protect human health. The reader is encouraged to look at the Tox21 website (tox21.gov) to get an idea of the type of effort that is needed to gain information on those thousands of chemicals. Tox21 is focused on testing mechanisms of toxicity in cellular or in vitro systems and examined the effects of 10,000 chemicals. Ultimately, we will have a great deal of mechanistic knowledge on these chemicals, but how do we integrate this information when we measure a random collection of 2187 different chemicals in an individual? Prioritizing and focusing on biologically active chemicals is an essential element of exposome research. Once we identify clusters of chemicals that exert particular biological effects, it will be necessary to have the external measures in the environment to identify the types of exposures and sources of those chemicals to engineer solutions to improve human health.

4.6 The microbes

Another major area of research that has found its way into the exposome domain is the microbiome. The microbiome consists of the millions of microbial organisms that exist within the confines of the human body.

Primary among these are those that reside in the intestines of the digestive system, but microbes also reside within our urogenital systems, the integumentary system, and also in our noses and mouths. These species include bacteria, viruses, and single-cell eukaryotes. The gut microflora has been somewhat underappreciated in the past, although most people acknowledge its importance after a course of strong antibiotics, which can decimate these important microorganisms. The human microbiome is thought to contain more than 10 times the total number of cells found in the human body, and due to their much smaller size (on a per cell level) they only contribute 1–5 lb of human body weight. With the diversity of organisms, there are likely 100 times more genes in the microbiome than in the human genome. There is potential for extensive metabolism of the nutrients and chemicals that enter our intestines. Harmful substances may be metabolized to less harmful species. Conversely, somewhat inert compounds may be bioactivated by the microbes to more harmful substances. These microbes can also influence absorption of nutrients and toxicants across the gut. This may be deleterious if nutrients are being blocked, but beneficial if harmful xenobiotics are being impeded. One of the advantages of microbiome research is that it completely builds upon the success of genomic technologies by simply sequencing the genomes of the microorganisms that reside within our bodies (discussed in Chapter 5, Measuring Exposures and their Impacts: practical and analytical). We will likely learn more about the role of the microbiome in health in the coming years. It is somewhat staggering that an appreciation of the role of the trillions of foreign cells in our bodies only developed about 20 years ago. It is possible that much of what we consider to be interindividual variability in response to exposures or drug treatments may be due to an individual's microbiome. Indeed, some conditions are even treated by repopulating the microbiome with specific microorganisms. It is important to consider the impact of the microbiome on the exposome. For example, an approach that only looks at the exogenous exposures without considering what happens to the chemicals once they get inside our body would totally miss the effects of the microbiome (Fig. 4.4).

Fortunately, the evaluation of chemicals in our bloodstream may somewhat obviate a need to specifically measure many aspects of the microbiome, since many of the processes in our bloodstream are occurring downstream from the effects of the microbiome. For example, if disruption of the microbiome reduced absorption of a particular nutrient by 50 percent, then one would expect to see much less of the nutrient in the bloodstream. It is possible that much of the variation that is observed among human metabolomes is the result of differential microbiomes. Such experiments could be readily conducted using humanized mouse models, in which the human microbiome is transferred into the mouse. An especially intriguing connection between the microbiome and the exposome is the recent work on the interface between the built environment and the microbiome (see work from the National Academy of Science http://nas-sites.org/builtmicrobiome/). Our exposures to microbes can be heavily influenced by our built environment. Regulation of humidity and the types of building materials can dictate our exposures to various microbes. Engineers who work on the built environment in the context of the microbiome have become critical contributors to exposome research as exemplified by the work of the laboratories of Amy Pruden, Peter Vikesland, Linsey Marr, and Marc Edwards at Virginia Tech (Dai et al., 2017).

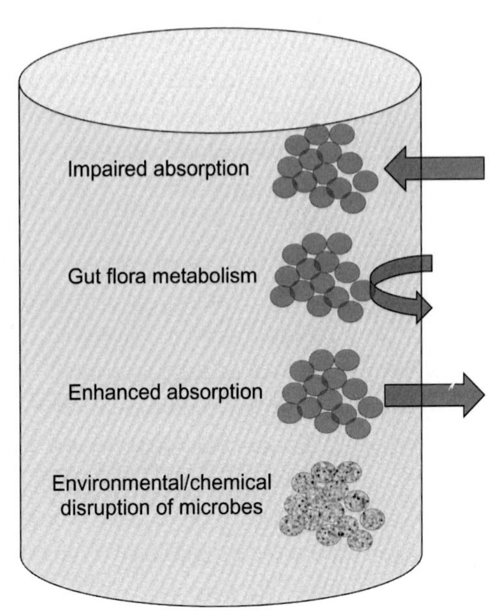

Figure 4.4 The microbiome–exposome interface. The microbes that reside in our digestive, urogenital, and integumentary systems have a significant impact on our health. As shown here, bacterial colonies located in the intestinal lumen can alter absorption of nutrients or toxicants from the gut and metabolize nutrients or toxicants (degrade or activate) before they are absorbed into our bloodstream, and environmental chemicals can disrupt the balance of the microbial colonies. The role of the microbiome in the body's response to environmental exposures is an important consideration for the exposome.

4.6.1 Sociology, society, and social determinants of health

One of the stronger offshoots of exposome research has come from the field of sociology, which focuses on the structure and function of human society with society representing the collection or population of humans. It captures group dynamic and social constructs. The term "socioexposome" has been coined (see Chapter 1, The Exposome: Purpose, Definitions, and Scope, again for my diatribe on balkanization of exposome). These investigators have interpreted the general definition of the exposome as the sum of exposures throughout the life course of the subject and how it can impact genes and gene expression. It is not surprising that much of the early research on the exposome took advantage of genomic tools. With that limited definition of the exposome, one can see how the lack of sociological constructs is glaring. Yet, the broader definition and interpretation outlined in the Miller and Jones' paper and in the first edition of this book are much more inclusive. I acknowledge that my use of the term "behavior" in our definition may be incomplete, but it was intended to capture nonchemical exposures-that is, activity, psychological influences, and personal interactions—all of which represent sociological concepts. If one considers behavior to mean human behavior and the resulting manifestations, which I intended, it should be inclusive of these ideas. The emphasis on the socioexposome brought welcome and critical expertise to exposome research.

Although my scientific lens is unapologetically biological and biochemical, I recognize that the most influential modifiers of health reside within the population as a whole. Prevention mechanisms afforded by the field of public health have the greatest potential to improve the health of the entire world. The social determinants of health are often more effective levers for improving human health than are the biomedical solutions(probably vastly more cost-effective on a population scale). Senier et al. suggested that without engagement and inclusion of sociological concepts, "exposome research could molecularize complex social phenomena, reducing the social experiences that condition population-level variations in exposome to individual-level molecular- level differences" (Senier, Brown, Shostak, & Hanna, 2017), further criticizing the reductionistic nature of exposome research. My simple response to this critique is "yes." The authors are correct. If exposome research takes on an ever more reductionistic approach, it risks being unable to address the underlying social conditions that drive exposures (it should be noted that Jones and I view the exposome as the antithesis and antidote to reductionism—it is a holistic scientific framework). However, the molecularization of social phenomena may not be a wholly bad thing. When social conditions lead to altered health outcomes, there must be a biological explanation for the altered health state. This is why Jones and I emphasized "... and the associated biological responses" in the definition of the exposome. The physiological impact and the body's response to such insults does underlie the health state. Steve Rappaport from Berkeley often referred to it as the idea that all environmental influences must be converted to a chemical signal in the body. If a social factor is impacting one's health, there should be a chemical/biochemical signal that drives it, and more important, that is measurable in biological samples. The ideal situation is to catalog the biochemical state occurring in the body along with the external factors that include social, climatic, policy, and all of the classic exposures measures. Senier et al. go on to state, "we believe that environmental health scientists should remain committed to the broad scope of the original exposome concept ... and that sociologists could be helpful and productive partners in this intellectual project," and I continue to agree with them. There is no doubt that one's sociological setting can be a major driver of health, and it could capture the primary cause of a subsequent epigenetic-driven health manifestation – for example, early traumatic event and future stress responses. It is important to note that this message of emphasizing sociological constructs is for more than just environmental health scientists, it is for anybody who adopts the exposome framework for their research, and my goal is for that to be a very large and almost grandiose tent. To capture these sociological concepts, it will be necessary for exposome research to use a population-scale approach, which will require a major scaling up of exposome-related activities. Senier et al. also suggest examining the exposome using an individual, local, and global framework, which is consistent with what is described in Chapter 10, The Exposome in the Future.

4.7 Obstacles and opportunities

Environmental health sciences is an umbrella discipline that is composed of multiple subdisciplines and is the natural home for the exposome. Within the field, communication among the population, exposure, and mechanistic scientists could be improved. Interaction among exposure scientists, toxicologists, and epidemiologists does not occur as often as one would hope. Enhanced collaboration among these areas is important for the health of the field, but it is especially important within the context of the exposome. The arbitrary boundaries that drive the reductionistic study of environmental factors in disease are eliminated, and their removal is likely responsible for some of the consternation from many scientists who work in the specific subdisciplines. Scientists have a tendency to be territorial and defend their scientific turf. While it is important to maintain the integrity of various scientific disciplines, science is moving and evolving at a rapid pace. Those poised to capitalize on the newest innovations within the context of their own discipline are likely to be in a stronger position in the future. Eventually, the defenders of scientific turf may look back to see that the grass is dying.

The broad field of environmental health sciences must be the driver of any exposome initiative, and given the critical need to reach out to numerous outside disciplines, it is essential to have a coordinated structure to entice these other fields to examine the problems surrounding the exposome. Currently, the exposome is being discussed within the various subdisciplines, but there has been little coordination among them. It will also be critical to recruit investigators from disciplines such as sociology, psychology, and engineering as we pursue the breadth of the exposome. This integration of multiple disciplines is an important obstacle to address. Indeed, for the outline provided in Chapter 10, The Exposome in the Future, to succeed, it must be driven by a unified environmental health sciences community with partnerships from many other fields. A key challenge is that departments of environmental health sciences almost always reside outside of medical schools, generally being stationed in schools of public health. Thus, it is incumbent upon the field of environmental health sciences to be the standard bearer of the exposome to the medical profession.

4.8 Discussion questions

Why is it important to address the mechanistic basis for associations between given exposures and health or disease outcomes?

Does the exposome add value to the field of environmental health sciences or is it a distraction?

What is outside the sphere of environmental health sciences? Where do the biological, physical, and chemical influences on health fit? What about behavioral, social, etc.?

References

Agathokleous, E., & Calabrese, E. J. (2019). Hormesis: The dose response for the 21st century: The future has arrived. *Toxicology, 425*, 152249. Available from https://doi.org/10.1016/j.tox.2019.152249.

Brunekreef, B. (2013). Exposure science, the exposome, and public health. *Environmental and Molecular Mutagenesis, 54*(7), 596–598. Available from https://doi.org/10.1002/em.21767.

Dai, D., Prussin, A. J., II, Marr, L. C., Vikesland, P. J., Edwards, M. A., & Pruden, A. (2017). Factors shaping the human exposome in the built environment: Opportunities for engineering control. *Environmental Science & Technology, 51*(14), 7759–7774. Available from https://doi.org/10.1021/acs.est.7b01097.

GBD. (2016). Global, regional, and national comparative risk assessment of 79 behavioural, environmental and occupational, and metabolic risks or clusters of risks, 1990–2015: A systematic analysis for the Global Burden of Disease Study 2015. *The Lancet, 388* (10053), 1659–1724. Available from https://doi.org/10.1016/S0140-6736(16)31679-8.

GBD. (2017). Global, regional, and national comparative risk assessment of 84 behavioural, environmental and occupational, and metabolic risks or clusters of risks, 1990–2016: A systematic analysis for the Global Burden of Disease Study 2016. *The Lancet, 390* (10100), 1345–1422. Available from https://doi.org/10.1016/S0140-6736(17)32366-8.

Lioy, P. J., & Rappaport, S. M. (2011). Exposure science and the exposome: An opportunity for coherence in the environmental health sciences. *Environmental Health Perspectives, 119*(11), A466–A467. Available from https://doi.org/10.1289/ehp.1104387.

Narahashi, T., & Haas, H. G. (1967). DDT: Interaction with nerve membrane conductance changes. *Science, 157*(3795), 1438–1440. Available from https://doi.org/10.1126/science.157.3795.1438.

Narahashi, T., Anderson, N. C., & Moore, J. W. (1966). Tetrodotoxin does not block excitation from inside the nerve membrane. *Science, 153*(3737), 765–767. Available from https://doi.org/10.1126/science.153.3737.765.

Rappaport, S. M., & Smith, M. T. (2010). Epidemiology. Environment and disease risks. *Science, 330*(6003), 460–461. Available from https://doi.org/10.1126/science.1192603.

Senier, L., Brown, P., Shostak, S., & Hanna, B. (2017). The socio-exposome: Advancing exposure science and environmental justice in a post-genomic era. *Environ Sociol., 3*(2), 107–121. Available from https://doi.org/10.1080/23251042.2016.1220848.

Further reading

Lioy, P. J. (2013). Exposure science: A need to focus on conducting scientific studies, rather than debating its concepts. *Journal of Exposure Science and Environmental Epidemiology, 23*, 455–456.

CHAPTER FIVE

Measuring exposures and their impacts: practical and analytical

A measurement is not an absolute thing, but only relates one entity to another.

Humphrey T. Pledge, Science since 1500: A short history of mathematics, physics, chemistry, biology. 1947

5.1 Introduction

This chapter focuses on how we currently measure the external forces covered in the previous chapter. Through our diet, breathing, and interaction with our surroundings, we are confronted with a myriad of exposures. These exposures cross our biological membranes in some manner and have the potential to impact our health. It is important to understand how these external forces alter our biology. This requires measuring the not only chemicals inside and outside of our body, but also the associated changes that occur when the chemicals interact with our biological components. In one of Wild's papers, he outlined the idea of the general and specific external exposome and the internal exposome (Wild, 2012)—the general being the environmental conditions that most people are exposed to and the specific referencing exposures an individual faces due to day-to-day activities. He then made the distinction with what is occurring inside the body, referring to it as the internal exposome. As noted earlier, I believe we should avoid fractionating the exposome or using modifiers to the exposome. I agree with Wild's organizational concept but disagree that they should be referred to as different exposomes. Obviously, how we measure what is in our atmosphere or water differs from how we measure chemicals in our body (to a degree). The matrices of air, water, plasma, and urine are different, but the technical approaches are converging. As Vermeulen, Schymanski, Barabasi, and Miller (2020) described in our recent article, high-resolution mass spectrometry

The Exposome.
DOI: https://doi.org/10.1016/B978-0-12-814079-6.00005-5

(HRMS) is emerging as the unifying technique to measure the exposome in environmental and biological media. Indeed, it has been the application of HRMS to the exposome that has allowed me to start using the term exposomics with confidence (a term I eschewed in the first version of this book). The exposome is angling to be the complement to the genome, which means we need people in other fields to appreciate this complementarity. It will be a nontrivial challenge to get other scientists to embrace the concept of the exposome. It is my belief that if we present multiple versions at the outset we will lose them.

This chapter provides an overview of the various approaches being used to measure components of the exposome. This is not meant to be comprehensive, as there are numerous reviews of the range of technologies, and I point the reader to them for further examination. It addresses the current state of the art in analytical chemistry, geotracking, satellite monitoring, and exposure science. It also examines the approaches we use to measure how environmental factors change our biology. I focus on the large-scale, omic-scale approaches. The growth in omic-based sciences to study biology has provided an enormous range of resources and techniques for measuring biology at a systems level (see Chapter 7: Pathways and Networks, and Chapter 8: Data Science and the Exposome, for more on systems biology and systems thinking). As noted in the opening of one of our recent papers:

> *Derived from the term exposure, the exposome is an omic-scale characterization of the nongenetic drivers of health and disease.*
>
> ***Niedzwiecki et al. (2019).***

The exposome provides a way to characterize exposures on an omic scale that is analogous to the tools we have for genomics, transcriptomics, epigenomics, and proteomics. Thus, I will focus on the approaches that allow the measurement of exposome components on an omic scale as these approaches will be the ones that are most likely to be incorporated into our omic-based biomedical research enterprise.

5.2 Exposure assessment

The field of exposure assessment had been making major contributions to exposome research well before the term was even coined.

Exposure assessors assess exposures. This field has developed approaches to measure contaminants in food, consumer products, air, water, and soil. Their work over decades created the foundation for the exposome. The field considers exposures that occur at the individual level and at the population level (similar to the general and specific concepts Wild used in describing the exposome). It is a useful framing, as it reflects the types of devices and sensors that may be necessary to identify these exposures.

At the population level, one can consider remote sensing using satellites, geographical information systems, and monitoring devices that work at the neighborhood, community, or regional level. Such systems can detect and monitor air pollution, water, radiation, noise, and the built environment (i.e., green spaces) at a broad scale. Particulate matter ($PM_{2.5}$) can be readily estimated at a global level using satellite-based technologies with optical imaging. When combined with sensitive mass spectrometry-based sensors placed in communities, it is possible to get detailed information on critical pollutants at a level of resolution that facilitates regulatory decision-making. This information must then be integrated into the dose and exposure modeling to determine what the conditions are in a particular area to predict possible exposures. At the individual level, one can use questionnaires or direct electronic feedback via smartphones, personal sensors, or biomonitoring. Biomonitoring is important for regulatory decision-making. By measuring the amount of chemicals in a person, one does get information on exposure and pharmacokinetics, but without complementary biological data, it is difficult to make direct connections between exposures and health outcomes. That said, biomonitoring is taking advantage of biological samples that generally do contain a wealth of biological information, and the study populations often have detailed outcomes data. I have had the pleasure of serving on the external advisory committee for the Human Biomonitoring for the European Union (HBM4EU; https://www.hbm4eu.eu/) for the past few years. This is a massive multicountry effort to obtain information on exposures across all of Europe. This coordinated project has helped strengthen infrastructure and advance the technology that is needed to monitor chemicals at a population level. It also illustrates the complexity of working with multiple governments and regulatory agencies to make decisions about prioritization and resource allocation. In biomonitoring programs, there are generally some specific chemicals or compounds of concern, and the focus is generally on these priority chemicals. Developing sensitive, robust, and reliable assays across multiple labs is essential. Yet, this approach relies on knowing about which chemicals one should be concerned. Coupling these targeted approaches with suspect screening (nontargeted) provides a mechanism for the discovery of unknown exposures and health risks within a biomonitoring program.

One interesting area of integration between individual- and population-level exposure assessments is the use of smartphones. The reader has certainly used a direction-finding map program on a smartphone. These systems take data from individuals to populate a multiuser map that can show traffic patterns—individual data to population-scale information. Since smartphones and activity monitors have the ability to measure other features such as heart rate, temperature, and light levels, there is potential to merge these data streams into programs that can measure responses to events or even anticipate larger health concerns. For example, we know that merchants see upticks in health remedy purchases that often precede health department knowledge of illnesses. Might there be similar observations made for exposures of chemicals or other environmental hazards? Being able to geocode an individual allows the scientist to compare that person's location to maps of exposure measured by monitoring stations, but it is unclear what level of resolution is needed. Do we need individual-level exposure? Household? Neighborhood? Zip code? County? State? It appears to depend on the type of exposures. For some pollutants, zip code–level information may suffice, for others, it may be necessary to have individual-level data.

Physiologically-based pharmacokinetic models are often used to estimate how much of a chemical a person is exposed to based on data from population-level exposures and individual-level exposure assessment. The idea is that if a person is in contact with a certain level of a pollutant, one can estimate how much of that chemical would cross through their skin or lungs and enter the body. The field of exposure science also considers the intermediate effects using a range of biomarkers. In this respect, there appears to be some overlap with the concept of the exposome, which is not unexpected since the exposome concept was initially built upon exposure assessment. However, addressing health effects or disease requires an epidemiological or clinical framework, and biomarkers alone do not capture the intermediate effects that represent the biological manifestations of exposures. This is where we need more integration. The current state of knowledge of the biology of disease or health can be overwhelming. One needs experts in those diseases and processes to get the full picture. The gap between the exposure assessors and the disease experts has been far too great. The exposome, as described in this book, is very much focused on this interface by providing technical tools and intellectual constructs, which allows one to bridge the gap.

One of the major challenges of integrating exposure assessment into clinical and translational research is that the resources needed for doing much of the environmental analysis exceed what is the perceived value of the information obtained from the analysis. If a clinical scientist focused on studying diabetes or depression can choose between imaging the pancreas or brain or having a suite of environmental exposures measured, they generally select the approach that fits their mental models. The researcher tends to go with the technique with which she or he is most familiar. This is why an exposomic analysis may be more palatable to the scientist not familiar with environmental health science. It is relatively easy for the investigator to layer in another omic technology because it fits within their framework of disease pathogenesis. An assessment of a patient's or a population's exposures with air monitors, occupational history, and chemical analysis does not fit within their intellectual framework. If one can demonstrate how complex exposure can be inserted into an existing multiomic framework, one can overcome some of the initial resistance. My colleagues and I have described this type of approach in two recent papers (Niedzwiecki et al., 2019; Vermeulen et al., 2020).

I am embracing my own bias here. I believe that HRMS is poised as the master technological integrator for exposure assessment, intermediate biomarkers of effect, and the metabolic features of health or disease state. By measuring chemical features in environmental media and in biological samples, HRMS can do it all. This is not to say that there are not complementary approaches and some limitations, but we have a single technology that gives us an unprecedented level of resolution and is poised to merge these often disparate sides of the environmental health coin (Vermeulen et al., 2020).

Chemicals, food systems, and detection

An example of the delicate balance between synthetic chemicals and human health can be seen in agriculture and food systems. Insecticides, herbicides, and fungicides, which are specifically designed to be toxic to their target, are used throughout the world to control insect, plant, and fungal pests that interfere with crop growth and spread disease. Many people are alive today

(*Continued*)

(Continued)

because of the use of these modern chemicals. At the same time, there is substantial evidence that many of these chemicals exert adverse effects on human health. Many of these studies were based on the ability to measure specific chemicals in human subjects. Total elimination of the use of these chemicals would likely cause famine and disease in many regions of the world. Thus, we must admit that these chemicals currently have a benefit to human health. At the same time, we also must acknowledge that many, if not most, of these chemicals have the potential to exert deleterious effects on human health and the ecosphere. Judicious use of these chemicals are essential and steps should be taken to use the safest and lowest amounts possible, including the use of integrated pest management strategies and the elimination of the use of chemicals when possible. Currently, the relationship with environmental health scientists and companies that manufacture and utilize these chemicals is mainly adversarial. Most chemical companies have focused more on their own corporate health than that of the general public, but with the proper motivation and regulation, these companies could develop practical and economically sound solutions to reduce the impact of their products on the environment.

The development of the field of green chemistry has been very beneficial to the chemical industry, but has had little impact on pesticide chemistry. Implementation of green chemistry ethos and approaches could help maintain control of pests and help reduce the negative impact on the environment. Synthesis of safer compounds with fewer toxic by-products could benefit all involved parties. Partnerships between chemical companies and environmental health scientists could accelerate this process. Indeed, the regulatory agencies across the world would be wise to implement strong incentives to companies to develop green chemistry approaches to pest control. The recent controversy over the restriction of chlorpyrifos is an interesting example. There is substantial evidence that exposure to chlorpyrifos has adverse health consequences. Its household use was restricted in the United States in 2000 (it was removed from the shelves of major home improvement chains), but it is still used in agriculture. The current U.S. administration has been criticized for not banning its use, but the prior administration had access to much of the same data and also did not take action. It is unfortunate that such important health decisions can be at the whim of politicians, but we have a bigger problem. Chlorpyrifos is one of dozens of organophosphate pesticides that share similar structures and actions. Are we prepared to ban them all? In the United States, we have options for insect control, but do other countries? Might a worldwide ban impact food supplies in poorer nations? What chemicals will replace the organophosphates? Pyrethroids? Neonicotinoids? Going chemical pesticide-

(*Continued*)

(Continued)
free is not a reasonable option across the globe (at least not with sufficient transition time for farmers, industries, regulators, and economies to adapt). Our regulatory bodies do not make decisions using a system-level approach. We play regulatory whack-a-mole, banning chemicals without evaluating the downstream effects of doing so. System-level approaches, such as those used in exposome research, may provide superior alternatives.

In addition to the expert environmental monitoring conducted by scientists, there is also an interesting community-level trend of personal monitoring. Part born out of interest in health and gadgetry and part from a citizen science perspective, more consumers are purchasing devices to monitor their personal environment. The increased awareness of one's environment is a positive trend, but the accuracy and actionable information from such sensors are questionable. Even so, the public is hungry for this sort of information. For example, the website quantifiedself.com focuses on how a person can measure everything possible about their health. The National Academy of Sciences held a workshop on personal monitoring that explored these topics in depth, and the meeting summary can be found here (http://nas-sites.org/emergingscience/meetings/personal-environmental-exposure-measurements-making-sense-and-making-use-of-emerging-capabilities/). Now that we have discussed what is going on outside of our bodies, let's explore some of the technologies that can tell us about what is happening inside our bodies.

5.3 Bandwagon science

Physicians have relied on physical examination and patient history for the diagnosis of disease for centuries. The last century has provided a wealth of biochemical measures that have enhanced the diagnostic ability of the medical field. From detecting altered amino acid metabolism, altered blood cholesterol, to identifying genetic mutations, the clinical laboratory has provided superb insight. In the 21st century, omics is making its way into the clinic. This represents a major shift from individual measures to system-level information, but there are major obstacles to be overcome before we have truly omic-scale data for personalized medicine.

Several fields have adopted the term omic to describe the large-scale approaches to their particular field (genomics, epigenomics, metabolomics, proteomics, etc.). A quick PubMed search for the various omes and omic technologies reveals a breadth of acceptance and utility. As noted earlier, the genome is obviously the most commonly used ome, and was coined nearly a century ago. Progress on the Human Genome Project spurred similar aspirations for areas examining gene transcription and the translation of proteins. A comprehensive analysis of the complement of genes, transcripts, and proteins was bound to occur. Similarly, understanding all of the ways that the genome could be modified via epigenetic mechanisms was another logical step. After these core -omes were described and corresponding projects initiated, several other fields jumped into the fray. Evaluating one's response to a drug, an exogenous chemical, based on their genomic profile gave birth to pharmacogenomics. Given the close relationship between pharmacology and toxicology, the development of toxicogenomics was another logical extension. More recently, the metabolome and microbiome have gained attention and recognition, along with a slew of other -omes, such as the connectome, which focuses on the connections among proteins, arguably a subset of the proteome; the lipidome (the profile of body lipids); the infectome (triggers of autoimmunity); and the phenome (set of all phenotypes within an organism or species). For dozens more examples, see http://omics.org/index.php/Omics_classification. Any effort to characterize our exposures at a similar scale will benefit from a close examination of many of these other -omes.

As given in Table. 5.1, the genome is the king of the -omes. A PubMed search of keywords highlights this. In 2013 the genome weighed in with 857,443 hits, proteome with 27,875, and transcriptome with 14,826. Pharmacogenomics (pharmacogenome is not a term that is employed) had 14,422 entries with metabolome at 3,167 and microbiome at 3,993. Epigenome had 1214 and epigenetics had 6,723. The connectome had 226 and phenome had 223. When one moves into the environmental health sphere, toxicogenomics comes in at 1,117, toxome at 4, and exposome with only 54. Interestingly, the word gene yielded 1,787,215, which is only twice that of genome. Transcript gave 58,146, approximately four times higher than transcriptome. Protein yielded 5,278,564, which is 189 times higher than proteome. Environment had 1,101,485, which is 20,000 times higher than exposome. What gives? Environment likely appears many times due to more general use of the

Table 5.1 Omes and citations
August 2013

-Ome (year coined)	Number of citations	Root term	Number of citations	Fold difference[c]
Genome (1920)	857,443	Gene	1,787,215	2.1
Proteome (1994)	27,875	Protein	5,278,564	189
Transcriptome (1997)	14,826	Transcript	58,146	3.9
Epigenomics (1950s)	1214	Epigenetics	6723	5.5
Toxicogenomics[a] (1999)	1117	Toxin	273,954	245.3
Exposome (2005)	54	Exposure[b]	583,775	10,811

A search of PubMed (2013) for the -ome term and its corresponding root word was conducted. The ratio between the -ome term and the root term gives a sense of the acceptance and use of the -ome term. Genome and proteome are well represented, while toxicogenomics and exposome are not. There is a strong denominator effect here in that there were only 54 citations of the work exposome as of July 2013.
[a]Toxicogenomics, like pharmacogenomics, never used the -ome suffix, only the -omic suffix.
[b]Substituting environment (1,101,485) for exposure yields a 20,398-fold differential.
[c]The right column shows the fold difference between the number of citations of the root term and the ome term.

Table 5.2 Omes and citations 6 years later.
December, 2019

-Ome (year coined)	Number of citations	Root term	Number of citations	Fold difference
Genome (1920)	1,470,774	Gene	2,685,539	1.8
Proteome (1994)	125,184	Protein	7,021,996	56
Transcriptome (1997)	86,276	Transcript	912,098	10.6
Epigenomics (1950s)	14,754	Epigenetics	89,674	6.1
Toxicogenomics (1999)	2,409	Toxin	366,552	152.2
Exposome (2005)	663	Exposure[#]	855,767	1291

The analysis of omic nomenclature was repeated 6 years after that conducted in Table 5.1. There has been substantial growth in all areas reflected the progress of science, although there are some areas that have seen more substantial growth.
[#]As indicated in Table 5.1 the word environment gives an even high number.

term environment, but still the amount of entries for exposome (the denominator) is paltry, likely a combination of its relative youth and slow adoption.

It is also interesting to examine how things have changed over the past six years (Table. 5.2). From this, we see that there is notable growth in the number of publications citing epigenomics and the exposome. Indeed, the growth of these two domains both of which capture alterations outside the encoded genome is higher than the classical omes of the central dogma. If one looks at the number of citations for each of the omes to examine the relative growth from 2012 to 2019, we see

Genome1.7 ×
Proteome4.5 ×
Transcriptome5.8 ×
Epigenomics12.2 ×
Toxicogenomics 2.2 ×
Exposome12.3 ×

The lower level of growth in toxicogenomics is notable, but the overall number of citations is still considerably higher than the exposome. Thus, while there appears to be a positive trend for the exposome, there is still room for improvement and growth. Next, the major omes will be summarized, highlighting some of the important technical approaches that have contributed to the discipline that allows us to examine how external factors impact our complex biology.

5.4 Genome/transcriptome

The ability to print thousands of nanoliter drops of complementary DNA (cDNA) onto a glass slide was a key step in the development of the microarray and the ability to measure thousands of transcripts in a single sample. Hybridization of mRNA to cDNA was a mainstay in molecular biology; it was just necessary to miniaturize the process. The laboratories of Pat Brown and David Botstein employed simple robotics and fountain pen technology to generate the first microarrays. After mRNA is isolated from a biological sample, a more stable sample of cDNA can be generated through the process of reverse transcription. By labeling two different populations with unique fluorophores, it was possible to measure the relative abundance of the bound sample, corresponding to a particular cDNA. Soon after, shorter oligonucleotides with overlapping sequences were spotted on the arrays that served as the basis of the Affymetrix platform. This technology avoided the need to keep large stocks of cDNAs, and because it used multiple overlapping oligonucleotides, it was able to detect changes at the single-nucleotide level. The cDNA approach did not have this level of specificity. The oligonucleotide-based approach also allowed the assessment of up to 2 million single-nucleotide polymorphisms (SNPs). As these approaches attempted to measure all of the transcripts, the results were often referred to as the transcriptome. Massively parallel sequencing, or next-generation sequencing, uses a variety of chemistries (pyrosequencing, reversible dye termination, and nanoballs), in combination with increasingly complex databases, to sequence DNA at ever-increasing speeds. More recently, these approaches have been used in the high-throughput sequencing of RNA (RNA-seq or whole-transcriptome shotgun sequencing). The potential benefits of RNA-seq are the more complete coverage of the more than 10 million SNPs in the human genome, the possibility of direct sequencing of mRNA without the need for reverse transcriptase, which can introduce errors in the sequence, and sequencing of non-mRNA species, such as noncoding RNA. The transcriptome will continue to advance as newer methods provide greater resolution and fidelity. Chapter 2, Genes, Genomes, and Genomics: Advances and Limitations, examined the importance of genomics, and the reader has access to hundreds of books and thousands of papers on the topic. The extraordinary advances in next-generation sequencing are making the prospect of full genomes a mere commodity.

5.5 Proteome

Proteomics has significant advantages over transcriptomics because it is measuring the actual effector molecules in the system—that is, proteins. One must infer that the altered gene transcription will ultimately lead to a change in protein expression, whereas proteomics is directly measuring the protein. Assessment of the entire complement of proteins in a cell or an organism is the goal of proteomics. Proteomics assesses which proteins are present, their relative abundance, and various modifications to the proteins. The separation of proteins based on size and charge is routinely performed using acrylamide gels. In fact, the initial proteomic studies were based on two-dimensional (2D) separation by size and isoelectric points. The various spots on the gel could be visualized by silver staining and the size of the points correlated to the relative abundance. The proteins could be transferred from the 2D gel to nitrocellulose and the been subjected to perform immunoblotting to identify proteins. In addition, by excising a particular spot and subjecting the purified sample to digestion and mass spectrometry, it was possible to identify proteins by the unique mass-to-charge ratio of the cleaved proteins using matrix-assisted laser desorption/ionization time-of-flight (MALDI-TOF) or electrospray ionization with time-of-flight (ESI-TOF) mass spectrometry. For example, a 70 kDa protein has about 60 amino acids. The trypsin digestion breaks it up into peptides with an average molecular mass of 1000 Da, about 9 to 10 amino acids. Thus, our sample protein is cleaved into six or seven pieces and that pattern of fragments, when compared with the reference database and reassembled computationally, allows identification of the protein. Even with a one-dimensional separation by mass, it is possible to distinguish among many proteins sharing the same total mass, as their trypsin fragmentation patterns are unique. The peptide fragmentation approach is laborious but this approach is an excellent way to identify proteins. However, it is not optimal for getting full coverage of the proteome. With refinement of the fundamental steps of the separation of proteins followed by identification by mass spectrometry, higher throughput technologies have become available.

Separation of tens of thousands of proteins in a given sample does not have to be conducted on an acrylamide gel. Ingenious protein chemists developed methods to separate the proteins with combinations of ESI or MALDI and tandem mass spectrometry (MS/MS because the mass spectrometry occurs in sequence with some sort of fragmentation in between) or Fourier-transform ion cyclotron resonance mass spectrometry (FTICR-MS). Instead of starting with cleaved peptides, the more recent approaches start with samples containing intact proteins. The proteins are fragmented during the process, but not with enzymes. These newer techniques are more expensive but more amenable to high-throughput proteomic experiments.

The multiple levels of modification and the temporal variability in expression make the proteome much more complicated than the stable genome. The identification of proteins is an important outcome of proteomics, but the ability of the proteomic approaches to identify protein modifications may be one of the field's strongest features. There are numerous ways proteins can be modified including glycosylation (carbohydrate/sugars), phosphorylation, ubiquitination (a marker for degradation), and sumoylation (small ubiquitin-like modifier that impacts localization and stability). These posttranslational modifications prime the protein for degradation via autophagic mechanisms (ubiquitination), activate enzyme systems (phosphorylation), and are involved in the maturation of proteins (glycosylation). Other forms of posttranslational modification include nitrosylation, oxidation, methylation, and acetylation. Oxidation was once thought to be a mere by-product of injury, but the reduced/oxidative state of proteins may serve a regulatory role for many proteins. Various initiatives to characterize the human proteome are underway. The Human Proteome Organization consists of two parallel projects. One is the Chromosomal-Based Human Proteome Project (C-HPP), with one research team assigned to each chromosome, and the other is the Biology/Disease-driven HPP (B/D-HPP) that will examine specific disease states and biological entities or processes. The Human Proteome Organization maintains a website that provides descriptions and updates (hupo.org).

5.6 Pharmacogenomics/toxicogenomics

There is evidence that people with certain genetic polymorphisms respond differently to certain drugs. Pharmacogenomics attempts to identify these sensitizing genes and then tests them in potential users. For the most part, these are specific genes that alter the susceptibility to a particular drug, either through enhanced or impaired metabolism or altered expression of the target molecule. There are many drugs that only work in individuals with a particular genetic profile. Many in the field believe that pharmacogenomics will become an integral part of personalized medicine by tailoring drug treatments to a person's genetic background. Toxicogenomics was conceived as a way to take advantage of microarray technology to measure the impact of toxic chemicals. Etymologically, it should have been toxico-transcriptomics, as it was using microarrays to assess the effects of toxic chemicals on gene expression. More recently, some in the field have proposed that toxicogenomics should include any omic data that assesses the toxicological consequences. If one accepts the term to encompass all omic technologies that inform the field of toxicology then it clearly falls under the larger umbrella of the exposome. This is analogous to the aforementioned Human Toxome Project (human-toxome.com) that is using omic technology to evaluate the toxic effects of chemicals. One could consider all of toxicogenomics and the toxome to be components of the exposome.

5.7 Epigenomics

By examining the epigenetic modifications discussed in Chapter 2, Genes, Genomes, and Genomics: Advances and Limitations, at a genome-wide scale, epigenomics provides a global measure of altered DNA regulation. In the United States, NIH has established a Human Epigenomics Mapping Consortium (HEMC) to serve as a resource for investigators. Taking advantage of next-generation sequencing and other advances, the consortium will generate and share data on DNA methylation, chromatin accessibility, histone modification, and small RNA transcripts at the genome-wide level. The ENCODE project, mentioned in Chapter 2, Genes, Genomes, and Genomics: Advances and Limitations, is serving as a repository for these epigenomic data sets. The HEMC now contains over 100 reference epigenomes, which raises an intriguing point. The first reference genomes were based on a few individuals. Of course, there is genetic variability, but the variability is only in the primary sequence. The epigenome, with its multiple levels of regulation, has much more variability. What is normal? What is the reference for the epigenome? Even without considering environmentally induced changes, one must acknowledge the changes that are part of development and specialization within cell types and organ systems. We will revisit the challenge of reference standards in later chapters as it relates to the exposome.

Another critical aspect of DNA methylation is its use in estimating biological age. Steve Horvath at UCLA noted that as organisms age that certain CpG sites are methylated. Since there appeared to be a propensity for some sites to be methylated more often than others, he devised an algorithm that focused on those sites that had the highest correlation to actual age. By analyzing the methylation status of these particular sites (a small subset, 353, of the millions of potential sites), he was able to provide a DNA methylation age that was proposed to correlate to biological aging (Horvath, 2013). This work has been met with a great deal of enthusiasm and has been integrated into the study of many diseases and conditions. As I raise in the closing chapter, the assessment of biological age via DNA methylation is precisely the type of biological consequence of interest to the exposome. Indeed, this could be one of the most important large-scale exposome studies to be pursued. A comparison of complex exposures to biological age represents a fundamental question in biology. That said, there are several challenges with using DNA methylation clocks, and the reader is referred to this excellent overview (Bell et al., 2019).

5.8 Metabolome

Metabolomics is the study of the chemical metabolites that result from cellular processes. Many of these metabolites are readily found in the bloodstream, providing an accessible source for sampling. By examining the state of the chemical processes within a cell or an organism, metabolomics provides information on the physiological state of the cell. For example, a transcriptomic or proteomic analysis may tell us that a particular enzymatic pathway is upregulated in a particular organ, but it does not tell us if the substrate is present, nor if the product is being formed. By providing information on the substrates and products, it is possible to determine what activities are occurring in the cells and quickly infer which proteins are active. Metabolomics shares many features of proteomics, in that one is attempting to measure the entire complement of the target entity (proteins vs metabolites) in a complex mixture. The challenges of separation and identification are quite similar. The major difference is that the metabolites tend to be smaller molecules than proteins. The molar mass of glucose is 180 g/mol, amino acids average about 110 g/mol, but proteins can be composed of hundreds to thousands of amino acids.

David Wishart at the University of Alberta reported the first draft of the human metabolome to contain 2500 metabolites, 1200 pharmaceuticals, and 3500 dietary constituents (see Human Metabolome Database, www.hmdb.ca). However, when one considers that plants also contain thousands of metabolites, the ingestion of a diverse diet greatly increases these numbers. It is also likely that many of the chemicals classified under pharmaceuticals are actually environmental contaminants. More recent work on the diversity in our diet reveals this complexity (see recommended reading Barabási, Menichetti, & Loscalzo, 2019). We tend to focus on just over 100 dietary constituents, when in fact there are more than 20,000 critical biochemicals in our diet, according to FoodDB. But this is limited to chemicals identified as being part of our nutritional biochemistry. When one considers the nonnutritive biochemicals in our food, the number grows even more. Similar to proteomics, a variety of mass spectrometry-based approaches have been used in metabolomics, including gas or liquid chromatography combined with mass spectrometry, tandem mass spectrometry, and FTICR-MS Nuclear magnetic resonance has also been used for metabolomic analysis. The goal is to be able to resolve as many of the metabolites as possible without sacrificing sensitivity. Advances in separation, detection, and bioinformatics will likely increase the sensitivity and resolution, leading to an even greater number of detected metabolites (see Fig. 5.1). I will refer to some of the key innovative steps in mass spectrometry in the next chapter, but if the reader is interested in learning more about the history of the use of mass spectrometry to conduct chemical analysis, I recommend this excellent review by from the Dorrenstein group (Aksenov, da Silva, Knight, Lopes, & Dorrestein, 2017). The unifying nature of HRMS merits a bit more attention. Although the source of the sample differs (environmental versus biological), the mass spectrometry, complex data extraction and processing, annotation to internal and external databases, and the suite of informatic approaches from uni- and multivariate statistics to network science are nearly identical. Measuring chemicals outside of our bodies in water, air, and food can use the same technology as measuring chemicals inside our bodies. The exposome provides a mechanism to bring together the environmental chemists with the scientists studying human diseases.

The breadth of technological options for metabolomics not only demonstrates a level of innovation, but also reveals some of the notable challenges. The development of mass spectrometers almost 100 years ago paved the way for the field. A major challenge was analyzing complex biological matrixes such as plasma or urine. A key first step is the chromatographic separation by liquid chromatography, gas chromatography, or capillary electrophoresis.

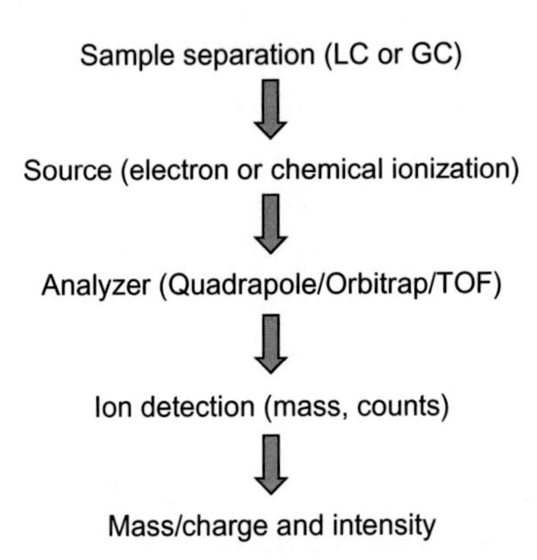

Figure 5.1 ***General workflow for metabolomics.*** Samples are typically separated by chromatography then introduced to the mass spectrometer, which consists of an ionizing source, a system to distinguish or separate the ions, and an ion detector. The electrochemical signal is then converted into a mass-to-charge ratio and includes the time at which it entered the source, which tells retention time.

The interaction with the various columns helps separate molecules based on size or charge before they are introduced to the mass spectrometer. The approaches use different stationary and mobile phases that provide different levels of resolution. For liquid chromatography the stationary phase could be an anion, cation, reversed phase, or hydrophilic interaction along with solvents like hexane, whereas in gas chromatography, a polymer lining the tubing serves as the stationary phase and inert gases are used for the mobile phase. Before introduction to the mass spectrometer the samples must be ionized. In gas chromatography mass spectrometry, the ionization occurs through electron (EI) or proton transfer (CI). Liquid chromatography mass spectrometry uses ESI or atmospheric pressure chemical ionization. MALDI provides a solid-state ion source.

After ionization the next step is the mass analysis. This is the heart of mass spectrometry. Quadrapole, TOF, ion trap, FTICR-MS, and Orbitrap are the primary forms of mass analysis. These techniques range in their resolution, accuracy, and sensitivity. A major challenge is that each approach has its strengths and weaknesses, and different laboratories have different objectives. This leads to a broad range of analytical approaches for metabolomics and a seemingly disorganized field. However, it is not disorganization, but rather a reflection of the level of technological innovation. That said, for large human studies like those needed for the exposome, there will need to be some convergence of certain techniques or common reporting.

From the laboratory standpoint, metabolomics provides the investigator with a list of features with mass-to-charge (*M/Z*) ratios and retention times (RT) for the various separations. There is additional information that can be gleaned by combining multiple levels of mass analysis, termed tandem mass spectrometry or MS/MS. Many instruments have multiple mass analyzers that allow the MS/MS to be performed (combined quadrapole, qTOF, and Orbitrap), and it is likely that this level of analysis will be essential for the ultimate identification of most molecules. For most M/Zs, there are dozens of potential matches. Unfortunately, our list of *M/Z*s and retention times do not tell us the identity of the molecules. Here, the field must rely on extensive databases to find matches, authentic standards to verify identify, and even additional preparative and structural analysis to characterize the features. Fortunately, several approaches are improving our ability to identify molecules. Some of these approaches are detailed in Chapter 7, Pathways and Networks.

I refer to our ability to identify features that come from metabolomic or exposomic studies more in Chapter 7, Pathways and Networks. The application of the techniques used in metabolomics to identify exogenous compounds can be referred to as exposomics. How does one identify a compound as exogenous? The first step is identifying the compound. This will likely be one of the greatest challenges for the exposome. Currently, metabolomics relies on annotation of features that come from the mass spectrometric analysis. Fig. 5.2 provides a schema for how one identifies molecules from HRMS. If one expects to obtain Level 1 structure by MS/MS and reference standard or by nuclear magnetic resonance identification, it will be a glacial process. This is not to say that there are not times for this level of rigor, but if we are trying to identify biological processes or pathways affected by chemical exposures, it may not be necessary to have Level 1 and 2 to have biological insight. There are several metabolomics databases available—for example, METLIN, MassBank, Human Metabolome Database, mzCloud, and GNPS, and matching unknown

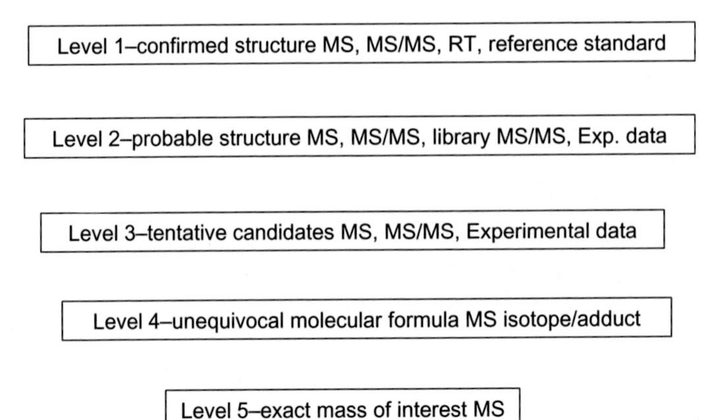

Figure 5.2 ***HRMS identifications.*** A major challenge for exposome research is the identification of compounds from HRMS. The schema shown here provides levels of confidence in identity. For the purist analytical chemist, only Level 1 will do, but when combined with other biological evidence, it is possible to generate useful data from lower levels of identification. For example, if one has Level 3 data on 8 of the 9 members of a particular pathway with one member confirmed with Level 2, it is reasonable to conclude that the biological pathway is worthy of further study. *HRMS*, High-resolution mass spectrometry. *MS*, Mass spectrometry. *RT*, Retention time. *This classification system is adapted from Schymanski, E. L., Jeon, J., Gulde, R., Fenner, K., Ruff, M., Singer, H. P., & Hollender, J. (2014). Identifying small molecules via high resolution mass spectrometry: Communicating confidence. Environmental Science & Technology, 48(4), 2097–2098. doi:10.1021/es5002105 (Schymanski et al., 2014).*

features with such libraries can help achieve Level 2. Tentative candidates can provide clues to biological effects and should not be summarily dismissed because the structural is not known with certainty. Knowing whether or not a compound is of exogenous origin may be impossible if it is naturally found in the body, but any halogenated compound can certainly be considered foreign. Well-controlled studies on large populations will help determine the range of normal endogenous features, one can expect to see. Repeated samples from subjects while monitoring their exposures may be the best way to ascertain whether or not a compound is of exogenous origin. Population-level information on workplace or geographical exposures should also reveal this.

5.9 What to do with all of these -omes

While it may be a bit premature to make this case, one could argue that the aforementioned -omes are actually all part of, contribute to, or are tools of the exposome. The microbiome was covered in the previous chapter, but all of those genes and metabolites are not part of the host-encoded genome and are readily classified as part of the exposome. If a person was placed into a vacuum, their genetic code (genome) would dictate the composition of their transcriptome and proteome, but once they exit the vacuum, these components start responding to the environment. The subsequent alterations in gene and protein expression, epigenetic changes, metabolic alterations, and changes in the microbiome all are part of the exposome. The responses are either genetically encoded or environmentally imbued. It is either genome or exposome (Chapter 3: Nurturing Science, Fig. 3.2).

5.10 The exposome and exposomics

I intentionally avoided using the term exposomics until 2017. Omics implies a unique approach. In general, omics refers to a particular field of study, while the ome refers to the object being studied. The exposome has been an entity hoping to be defined. Exposures, i.e. the exposome, are the object of study, but the techniques and approaches being used to study the exposures and associated biological effects already exist in a variety of fields. When I wrote the first edition of this book, I argued that no new techniques were being used to define the exposome; thus, there may not be a need for a new field of exposomics. In fact, I argued that environmental health sciences was the field primarily studying the exposome and that the exposome is just one particular component of that field. I was concerned that using a term that was not definable could be problematic. I argued that a human exposome project would require advancement in technology, but this could be an advancement in LC–MS, bioinformatics, or computational biology; technologies that are not exclusive to the exposome. The integration of the myriad of data from existing disciplines may require some unique approaches, but the study of complex biological processes and systems is referred to as systems biology (more on this in Chapter 8, Data Science and the Exposome). Five years ago, I preferred to view the exposome as the goal, not the process. Oh, how times have changed. Efforts led by the U.S. EPA and instrumentation advances have created some approaches that are decidedly exposome-friendly. Much of this work was described in the recent paper in Science (Vermeulen et al, 2020). In that article, we describe an experimental framework that includes study design, analytical chemistry, bioinformatics, and data science that together represents what I confidently refer to as exposomics. I believe the continuing advances in HRMS will help further define exposomics. Given the fact that the same approaches are being applied to measure environmental media further illustrates is foundational role in exposomics. I discuss this example in more depth in the following chapter on Innovation.

What Wild pointed out when he coined the term exposome was that the epidemiologists were not able to embed all of the exposure assessors' measures into their epidemiological framework the way that genetics had been able to do. Everybody knew that the environmental factors were important, but how could one link the external forces with biological alterations and health outcomes? The lull between Wild's 2005 paper and the upturn in related publications had much to do with the fact that there had not been major advances that made exposome research possible. In 2020 a combination of technological advances, specific application of those advances by innovative scientists, and improved bioinformatic and cheminformatic platforms has created the basis of exposomics.

5.11 Obstacles and opportunities

With unlimited resources, we could measure many environmental contaminants in our cities and neighborhoods, but we do not have unlimited resources. A challenge for the exposure scientist is to determine what the optimal cost–benefit ratio for the range of resolutions is. Which exposures do we need individual data on and which exposures will zip code–level data suffice? What is the cost of getting to the level of resolution to study disease outcomes? How much should we invest in individual sensors and monitors? Do they add value or merely add unnecessary cost and data?

From the biological side, we are seeing major advances in omic-level technologies. These approaches are being developed and optimized for other fields. Exposome investigators can select from these existing technologies to examine the biological effects of exposures. We do need to keep exposome research at the forefront of the research and development teams in the industry. We need continued advancement, especially on the environmental front. Most pharmaceutical drugs are in the databases for matching detected features, but their metabolites are not. This is surprising given the obvious interest the pharmaceutical industry has in such information. It is also notable that we usually know how much was introduced into the body. Turn toward the environment where we are typically looking at chemicals that were not intended to be there. The dose is poorly understood, and we have little understanding of how these chemicals interact with each other and with chemical and biological compounds inside our bodies (including microbial proteins and metabolic machinery). We need dedicated effort to get these chemicals and by-products into the databases. Without that information, we are left with many clues but few solutions.

I close this chapter with a warning about adopting -omic vocabulary. Over 100 omic terms have been coined, including the intentionally absurd ridiculome (http://www.biomedcentral.com/1741-7007/10/92). This omic abuse has, not surprisingly, led to the creation of the hashtag badomics (#badomics). The explosion of omes and omic terms has been somewhat unfortunate in the biomedical sciences. The idea of a comprehensive evaluation of certain biological data sets is admirable, but coining terms to artificially boost the apparent importance of one's topic of study is not desirable. It is critical for investigators in environmental health sciences to avoid such a perceived fate for the exposome. The field has struggled with how to deal with complex exposures, especially as it relates to disease causation/contribution and the development of regulatory guidelines. While many have criticized the concept of the exposome (or at least the word), few would summarily dismiss the idea that the environment has an impact on human health, and that these exposures represent complex mixtures. Thus, we must be methodical and careful in our approach to define the exposome. We must focus on verifiable outcomes that have the potential to impact human health. We also want to take advantage of the data and tools resulting from other relevant and validated omic-based technologies. While comparisons with the genome, transcriptome, and proteome are welcome, we would do best to avoid comparison with the ridiculome.

5.12 Discussion questions

Refer back to the quote that opened this chapter. Are there absolutes? Does the exact concentration of a chemical in our body matter? Is it all relative? If so, in relation to what? Endogenous chemicals? Unified standards?

If a study of the human genome gave rise to genomics, does the human exposome give rise to a new field of exposomics? Should it be a separate field or should the exposome serve as an organizing framework for the cornucopia of -omic technologies used within environmental health?

Should the exposome field focus exclusively on measuring chemicals derived from exogenous sources or should the field also evaluate the chemicals generated and processed from endogenous sources?

References

Aksenov, A. A., da Silva, R., Knight, R., Lopes, N. P., & Dorrestein, P. C. (2017). Global chemical analysis of biology by mass spectrometry. *Nature Reviews Chemistry*, *1*, 0054.

Barabási, A., Menichetti, G., & Loscalzo, J. (2019). The unmapped chemical complexity of our diet. *Nature Food*. Available from https://doi.org/10.1038/s43016-019-0005-1.

Bell, C. G., Lowe, R., Adams, P. D., et al. (2019). DNA methylation aging clocks: challenges and recommendations. *Genome Biology*, *20*, 249. Available from https://doi.org/10.1186/s13059-019-1824-y.

Horvath, S. (2013). DNA methylation age of human tissues and cell types. *Genome Biology*, *14*, 3156. Available from https://doi.org/10.1186/gb-2013-14-10-r115.

Niedzwiecki, M. M., Walker, D. I., Vermeulen, R., Chadeau-Hyam, M., Jones, D. P., & Miller, G. W. (2019). The exposome: Molecules to populations. *Annual Reviews of Pharmacology and Toxicology*, *59*, 107–127. Available from https://doi:10.1146/annurev-pharmtox-010818-021315.

Schymanski, E. L., Jeon, J., Gulde, R., Fenner, K., Ruff, M., Singer, H. P., & Hollender, J. (2014). Identifying small molecules via high resolution mass spectrometry: Communicating confidence. *Environmental Science & Technology*, *48*(4), 2097–2098. Available from https://doi.org/10.1021/es5002105.

Vermeulen, R., Schymanski, E. L., Barabási, A.-L., & Miller, G. W. (2020). The exposome and health: Where chemistry meets biology. *Science*, *367*, 392–396.

Wild, C. P. (2012). The exposome: From concept to utility. *International Journal of Epidemiology*, *41*(1), 24–32. Available from https://doi.org/10.1093/ije/dyr236.

Further reading

Aebersold, R., Bader, G. D., Edwards, A. M., van Eyk, J. E., Kussmann, M., & Qin, J. (2013). The biology/disease-driven human proteome project (B/D-HPP): Enabling protein research for the life sciences community. *Journal of Proteome Research*, *12*, 23–27.

Bloszies, C. S., & Fiehn, O. (2018). Using untargeted metabolomics for detecting exposome compounds. *Current Opinion in Toxicology*, *8*, 87–92.

Caron, H., van Schaik, B., van der Mee, M., Baas, F., Riggins, G., & van Sluis, P. (2001). The human transcriptome map: Clustering of highly expressed genes in chromosomal domains, 2001 *Science*, *291*(5507), 1289–1292.

CRC Press, Taylor & Francis Group. (2020). *Metabolomics: Practical guide to design and analysis*. Boca Raton, FL: CRC Press, Taylor & Francis Group.

The Human Microbiome Project Consortium. (2012). The Human Microbiome Project Consortium, Structure, function and diversity of the healthy human microbiome. *Nature*, *486*(2012), 207–214.

Walker, D. I., Valvi, D., Rothman, N., Lan, Q., Miller, G. W., & Jones, D. P. (2019). The Metabolome: A key measure for exposome research in epidemiology. *Current Epidemiology Reports*, *6*, 93–103.

CHAPTER SIX

Innovation and the exposome

To invent, you need a good imagination and a pile of junk.

Thomas A. Edison

6.1 Introduction

One of the responsibilities of my current and past academic positions has been to encourage and foster innovation. Although there are not ready-made training courses for those charged with leading innovation, there are numerous stories, studies, and opinions that can be drawn upon as one develops templates for innovation. Unlike the well-controlled laboratory from which many of us come, it is rather difficult to evaluate the success of these efforts. That said, this chapter builds upon Chapter 5, Measuring Exposures and their Impacts: Practical and Analytical, and wades into this murky sea of information/disinformation in an attempt to identify some strategies that may help foster innovation and propel the exposome field forward.

What exactly does it mean to be innovative? How does a new field position itself for innovation? What approaches or techniques will most benefit from innovation and novel solutions? How has innovation advanced other fields? Where should future efforts be focused? It is relatively easy to think about how improvements in laboratory equipment and techniques can help drive a field forward. Many Nobel prizes have been awarded for these technological breakthroughs, including cryo-electron microscopy, PCR, and DNA sequencing. The exposome would benefit from similar technological advances, but the field will need more than improvements in equipment to do what will provide the comprehensive analysis of our complex exposures that is sorely needed. Improved approaches for study design, data storage, and analyses will be just as important. Moreover, given the complexity of the problem, the data, and the analyses, the mindset of the exposome scientist may need to be a bit

The Exposome.
DOI: https://doi.org/10.1016/B978-0-12-814079-6.00006-7

different than that of an environmental epidemiologist or the mechanistic toxicologist. The exposome will need innovation in technology, analysis, and thinking to develop the analytical and computational platforms that permit large-scale assessment of the external forces acting upon us.

6.2 A pile of something

When Edison stated that "to invent, you need a good imagination and a pile of junk" he was being quite literal in describing a pile of junk— scraps of wire, metal parts, broken equipment, and old tools— as being all that he needed to create many of his inventions. I suspect that the engineer attempting to make a better mass spectrometer has a substantial number of parts and scraps from previous efforts. Given that high-resolution mass spectrometry (HRMS) has become the de facto machinery of exposome research, we will need engineers combining imagination and piles of technologically advanced junk to invent the ideal instrumentation for the exposome. Intronic DNA was once referred to as junk and is now increasingly being recognized as potentially useful and important. I think the exposome field has pieces in its "pile of junk" that are similar to what geneticists faced. Low-level exposures, often not high enough to cause overt toxicity, could be considered analogous to "irrelevant" junk DNA. The noisy background contaminants that are typically ignored are more likely than not having a deleterious effect on our health, albeit in combination with each other. Thus, with a dash of good imagination and ingenuity, these low-abundance chemicals and unidentified transformation products once thought to be insignificant will be revealed to be important environmental exposures as part of the exposome.

6.3 Moonshots and missed shots

In 2016 Vice President Joseph Biden helped launch the Cancer Moonshot in the U.S. Biden and a large number of senators were, and are, well into their 60s or 70s, so they remember when no person had been to the moon and are inspired by the term "moonshot." The goal of sending a person to the moon was just about as great of a scientific feat one could imagine. They remember watching the moon landing in black and white. Color television had been introduced and networks had color programming, but U.S. National Aeronautic Space Agency (NASA) cameras were black and white. We live in societies with advanced technological settings, and unfortunately, this results in the moon landing seeming passé. For a current college student the lunar landing is ancient history. It is something their parents (or grandparents) talk about. Black and white television or video is incomprehensable. Can you imagine asking a university student to binge-watch their favorite show on a grainy black and white TV? Nothing about it seems especially innovative except that it may make people think about the 2015 movie "The Martian" depicting humans living on Mars. University students cannot fathom the state of computers at the time of the moon landing. We are not even talking about floppy disks (the anachronistic save icons on our computers show an image of the rigid and comparatively modern 3.5 in. disk that few people in their 20s have ever seen). The instruments used paper punch cards. We did not have cellular telephones when they launched the Apollo rocket toward the moon. We were still tethered to the kitchen phone by a stretchy 8-foot phone cord. The disconnect between the branding of an ambitious scientific project and an event that occurred over 50 years ago is important. I do not believe that the majority of people under 50 years of age truly understand and appreciate what a moonshot is. New ideas and discoveries are going to be primarily made by generations that do not even understand the analogy. One must go back to the mindset of the 1960s to appreciate how truly astonishing the feat was. Let us examine what was happening in the 1960s that led up to this extraordinary event to help illustrate the importance of this innovation and the unfortunate disconnect with today's generation.

In 1969 NASA launched the Apollo 11 rocket toward the moon. Neil Armstrong was the first human to step upon the surface of the moon. In the 1960s the Soviet Union (U.S.S.R.) and the U.S. were in an arms race. They were both developing nuclear weapons at breakneck pace (with truly extraordinary advances in physics and engineering). Most of the efforts were conducted in secret with the general public knowing little about the specific advances, but not so for the race to the moon. The battle for technological superiority was on full display. Rockets were being launched into space—initially with no personnel, then with test animals, and eventually with humans. The U.S. had its Pioneer, Ranger, Mercury,

Gemini, and Apollo series rockets while the U.S.S.R. had its Vostok, Voskhod, Luna, and Soyuz series. We recall the successful landings and explorations but do not remember the failures...and there were many. Hollywood makes movies about the Apollo series, but not about Pioneer and Ranger, where failure to reach the moon was the norm. The 1957 Soviet Sputnik I was the first engineered orbiting satellite. In another first for the Soviets, it was the Soviet Luna craft that first reached the surface of the moon. Soviet pilot Yuri Gagarin was also the first human in outer space. The Soviets were leading the way. Yet, the singular event of landing and walking on the surface of the moon by the Apollo astronauts, considered a triumph of innovation, gets most of the recognition. It did not occur in isolation. There were parallel stories of innovation occurring on two continents.

The technological acumen of the Soviets and Americans was not that different. Both were delivering extraordinary advances, and the Soviets were ahead at many steps. Although the collapse of the U.S.S.R. has allowed access to more documentation, there is less information available about the setting that permitted this grand level of innovation in the Soviet setting. The U.S. efforts have been chronicled in books and movies and are literally on full display at museums. NASA had substantial resources, a clear set of goals that included intermediate steps, and support from superiors and constituents. The intellectual and managerial resources for the project were superb. Some of the most brilliant scientists in the world were fully dedicated to the project. Failure was permissible, but incompetence was not. These are all important lessons for those who desire innovation. The end result is the compilation of a series of innovative leaps that continue to build upon one another. The process that permitted sustained creativity and discipline is as important as the individual scientific and engineering feats. There is no doubt that the lunar landing that includes humans walking on the surface of the moon was one of the most remarkable scientific accomplishments of all time, but unfortunately, it is so far removed from our current day that the term "moonshot" is tepid. Let us come back down to the Earth to reconsider the exposome and the technological leaps that will be needed to achieve the full vision of the exposome.

The exposome represents a scientific construct that requires a reimagination of how we assess environmental influences on health. It requires a combination of new thinking and new technology. A human exposome project, as I discuss in more detail in the closing chapters, is big science. It should model itself after some of the most successful and daring scientific projects to date (Human Genome Project, Hadron Accelerator, development of computer microprocessors). The field must think far outside of its comfort zone.

6.4 Great by association

How does one create a scientific culture of innovation? There are many books written about innovation. Most are focused on the business community and have questionable methods and analyses. Generally, these books are a series of loosely collected and poorly validated anecdotes. They describe the conditions that were present when companies were succeeding, but are unable to identify causal relationships. There is a considerable amount of confirmation bias and little scientific rigor. For example, in the highly touted book *Good to Great* by Jim Collins, the author highlights several companies that went from being solid companies to being outstanding companies. He compared these high-performing companies with peer companies that did not elevate their performance. His team poured through the literature, interviewed many CEOs, and analyzed the data. It all seemed reasonable. Compare these two groups and see what traits differentiated their performances. It seems analogous to how we compare groups of healthy individuals with groups of individuals with a particular health condition or disease. Unfortunately, the traits identified that track a high-performing company are not necessarily causative. In *The Halo Effect,* Phil Rosenzweig went back and examined those good-to-great companies and conducted a rigorous statistical analysis with additional follow-up timepoints and a much larger company sample size. Rosenzweig revealed a major flaw because the author of *Good to Great* started by selecting companies based on outcome and collecting retrospective (and certainly biased) data. As the title suggests, these traits that seem to make good companies become great are in reality the traits that one observes when a company is growing. A company that is performing well, for whatever reason, will exhibit certain characteristics due to that growth. Thus the approach must control for performance. When you are working for a company that is hiring new people, there is a certain level of energy from the infusion of new talent. The converse is also true. When companies are cutting back, morale is going to be low. As companies grow to unprecedented sizes, do any of the predictors of performance that were based on smaller organizations decades earlier hold true? Thus, many of the traits extolled in business books as contributing to success are simply a reflection of the performance itself. Correlation, not causation. This is why the exposome needs well-designed prospective studies that avoid the identification of halos or clouds hanging over the heads of people with certain health conditions. For example, inflammation is common in many disorders, so one would see numerous biomarkers of inflammation in a diseased population, but this does not mean that the inflammation was involved in the underlying pathogenesis.

6.5 Acceptable failure, intolerable incompetence

In the first edition of this book, I noted my disdain for consensus (more on that later). When a group can achieve consensus, it generally means that the most innovative ideas are being suppressed. Recently, I came across the work of Gary Pisano in the *Harvard Business Review,* which captured the essence of my concern. He stated,

> *consensus is poison for rapid decision making and navigating the complex problems associated with transformational innovation.*
>
> ***Pisano (2019)***

His use of the word "poison" delights the toxicologist in me. Indeed, the processes needed for transformational innovation do not lend themselves to groupthink and consensus. He goes on to point out some of the features of innovation. It is notable that although his work is primarily focused on the business sector and has some of the same limitations mentioned previously, the research methods are stronger and the recommendations appear to comport with my own scientific thinking. Thus, I believe these ideas can be reasonably applied to the exposome and biomedical science. Fig. 6.1 shows an example of how the process of innovation goes through multiple, iterative steps.

One of the critical features of innovation is the difference between failure and incompetence (Pisano, 2019). Failed attempts at any endeavor can provide actionable feedback and increase the probability of the next attempt to succeed, or they can just reinforce the bad practices that led to failure in the first place. The former is acceptable failure, the latter is

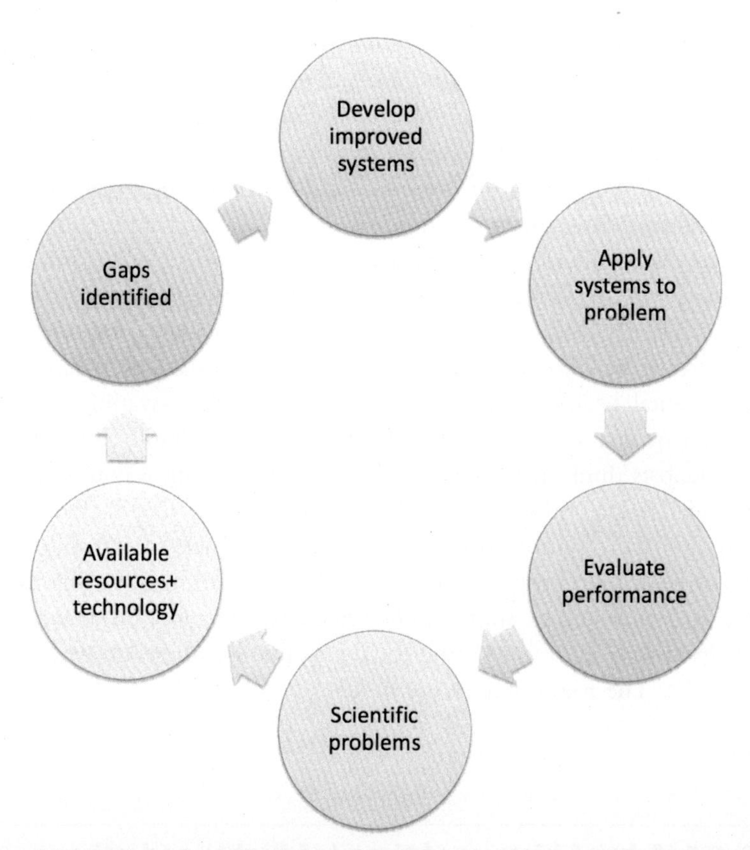

Figure 6.1 ***Iterative science for innovation.*** Innovation is an iterative process. First, problems should be identified. Then the existing technologies and approaches are examined for their ability to solve the problem. Gaps or needs for improvement are identified and guide future development. Those improved systems are then reapplied to the problem. Then a detailed analysis of the success or failure of the improvement is conducted. Then the original problem is revisited to see if the new approach can solve the original problem or even expand the original problem because of enhanced capabilities.

unacceptable incompetence. Innovative organizations must embrace the failure, but reject the incompetence. This can be a difficult balance to achieve. We want to believe that failed attempts will lead to better outcomes, but this occurs only when accurate data about the basis of the failure are captured, examined, and accepted. Even then it requires a range of adjustments to increase the likelihood of future success. This means hiring people with the right energy and creativity for the particular job and if it is clear that they are not up to the task, dismissing them as soon as possible. It is be much easier to make personnel decisions in the board room than it is in academia. In academia tenure and long-term appointments can prevent rapid change, but at the same time that security can be an advantage by providing a level of security to tinker and innovate.

As research progresses on the exposome, we must be painfully honest about failure versus incompetence. Carefully planned and meticulously executed experiments yield great knowledge. Poorly planned or poorly conducted experiments and research programs, or describing or branding work as the exposome when it is not working on an omic level, will detract from the efforts. The exposome is not just a new moniker for environmental health. It requires a daring level of innovation combined with high-quality research. Most scientists are not willfully incompetent. Most scientists think that the work they are conducting is contributing to a particular field. Unfortunately, there is some work that is poorly conducted due to a combination of flawed design, lack of statistical rigor, poor reproducibility, tepid creativity, and confirmation bias. This is why the field of exposomics must establish standards for the type of work that will help build a solid foundation for the future and these are described in Chapter 10, The Exposome in the Future.

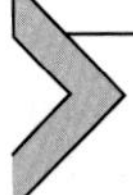

6.6 A brief history of time (of flight) and other technologies

The basics of high-resolution mass spectrometry (HRMS) were reviewed in the previous chapter, so this history will be very brief. The heading title, borrowed from the book *A Brief History of Time* by Stephen Hawking in a chapter about innovation, is apropos. Hawking challenged many of the long-held beliefs in physics and cosmology and sparked new fields of study. When one looks at the advances in chemical analysis, the innovative advancements in HRMS stand out as the essential features. Many of these technological advances resulted from academic research, but all required industry partnership to make them a practical reality. Moreover, many of these advancements occurred outside of academia.

HRMS has been had an impressive history of innovation. The first solid and electron impact sources were developed 100 years ago, the time of flight mass analyzer was invented in 1946, and the quadrapole mass filter and ion trap were introduced in 1953. The next few decades saw chemical ionization, field desorption ionization, and the advent of triple-quadropole mass analyzers. The work of Alan Marshall and Melvin Comisarow introduced the Fourier transform approach to ion cyclotron resonance. Marshall and Comisarow trained at Stanford, and Marshall focused on nuclear magnetic resonance (NMR) initially, with Comisarow focusing on ion cyclotron resonance. Their combined expertise came together to create the Fourier transform ion cyclotron resonance (FT-ICRMS). In the late 1980s electrospray ionization and matrix-assisted laser desorption/ionization appeared and revolutionized the analysis of biological samples. Although some of the technology behind the Orbitrap instruments had existed for nearly 100 years, it was not until the 2000s that the Orbitrap instruments became a commercial reality, providing high resolution and reliability. These innovations were driven by a desired to advance HRMS and metabolomics, not to advance exposome research. That has just been a happy by-product. A major goal of exposome investigators is to get the companies that make high-resolution mass spectrometers and their associated equipment to dedicate research and development on identifying and quantifying exogenous components. The technological development driven by genomics was beautifully extolled in the book *The Genome Wars* (Shreeve, 2004). Companies such as Applied Biosystems, Human Genome Sciences, and Celera were created or reformulated to help solve the problem, building upon the creativity of pioneers like Leroy Hood. The exposome will need a similar level of industrial innovation over the next 10 to 15 years to achieve the level of coverage that we need.

The application of these technologies to the exposome is still in its infancy. Dean Jones at Emory University has been applying the Thermo Orbitrap technology for exposome-related projects for several years. More recently, a handful of top mass spectrometrists have started to expand into areas that support the exposome by improving the identification of exogenous and previously unidentified features, which is precisely the type of innovation and effort that is needed. There has been outstanding work by Oliver Fein and Dinesh Barupal at the University of California at Davis, Gary Siuzdak, Minglang Fang (now at Nanyang Technical University), Caroline Johnson (now at Yale University), Benedikt Warth (now at University of Vienna) at the Scripps Institute, Emma Schymanski at the University of Luxembourg, and seminal work at David Wishart's group at the University of Alberta. The attraction of previously nonexposome, nonenvironmental health scientists into the exposome realm suggests that the exposome is increasing its visibility as "something important to be clever about" as further described below.

One of the underappreciated partners in exposome research is the companies that design and make the equipment we are using to measure exposures and responses. I am convinced that for the human exposome to reach its potential, we must have industry partners that recognize the long-term need, demonstrate a willingness to invest in technology development that will benefit the exposome field, and are viewed as intellectual partners in this important endeavor. We know that Agilent and ThermoFisher have focused on some efforts in this area, but will SCIEX, Waters, Bruker, PerkinElmer, Hitachi, and Shimadzu get into the game? How do you get companies to dedicate effort to develop new technology? How do you create competition among these companies to accelerate technological development? Since companies are in business to make money, it is necessary for there to be a financial incentive. If countries across Asia, Europe, and the Americas decided to run exposome-scale studies across their populations, there would be a substantial market for the instrumentation needed to perform the analyses. To date, there has not been an organized effort to motivate these corporations to contribute to the innovation we need for exposome research.

How do we wrap exposome research in a sheath of intellectual myelin that will permit innovation similar to saltatory conduction? As many will recall from the biology of neural conduction, the myelin sheath that envelopes many axons provides insulation that allows the action potential to jump from one node of Ranvier to the next (imagine running on the moon with giant leaps). This is referred to as saltatory conduction and provides an excellent analogy for leaps of innovation. The paradigmatic shifts in science referred to in Chapter 1, The Exposome: Purpose, Definitions, and Scope, could be considered to be great saltatory leaps, and the exposome will need similar advances over the next 10 to 20 years.

6.7 Measuring the previously unmeasurable

The Human Genome Project was an exemplar of scientific advancement and innovation. From dramatic increases in DNA sequencing to the computer algorithms used to assemble the segments of DNA, the Human Genome Project exuded innovation. The technological innovation for genome research was fit for purpose and it was a grand purpose. It was clear that there was a major need for sequencing technology. As noted in Chapter 2, Genes, Genomes, and Genomics: Advances and Limitations, the project was accelerated due to some robust competition from Venter's team, which is characterized in *The Genome War* by Shreeve. The external competition from Venter's Celera Genomics pushed the government-based Human Genome Project to become more creative. Venter's group had decided the ABI's Prism 3700 would provide the speed that they needed and were ordering hundreds of the instruments. Mike Hunkapiller was leading the ABI development, and the technology and instrumentation was so new to the production that it was not clear if they could keep up with the dramatic increase in orders, but the scaling was relatively simple for the genome efforts. If each instrument could sequence X bases per day, it was simple arithmetic to calculate the throughput. Eric Lander of the Whitehead Institute recognized the need to accelerate the NIH effort and was able to marshal support from his own institute to raise the stakes. The three major NIH genome centers converged on ABI's new Prism 3700 technology. They set ambitious goals and enlisted the help of many other genome centers across the world.

What is the analogous innovation for the exposome? Will it be future advances in high-resolution mass spectrometry (HRMS)? HRMS has been the beneficiary of several major innovations; without them I do not think I would be writing this book. I predict that advances in HRMS that are tailored to the needs of exposome research (separation, detection, databases, bioinformatics) will be recognized as essential elements of success. If an international consortium decided that they needed an instrument that increased resolution and decreased cost to a certain level and could convince the industries that several hundred orders would come in, the probability of technological leaps would increase dramatically.

Exposure scientists and toxicologists often cite the increased sensitivity in measuring environmental chemicals as evidence of scientific advancement. Older systems could measure chemicals on a micromolar scale. New instruments and technologies are allowing us to measure many chemicals on femtomolar levels. Thus we have excellent evidence that technological advancement improves our ability to measure. In fact, many argue that our existing technologies can do much of what we need to do for the exposome. It is apparent that we are very good at measuring what we know we want to measure. The problem is that the exposome embraces the idea that we do not know what we want to measure. We need techniques to measure thousands of known and unknown chemicals at the same time in biological or environmental samples. We must achieve an environment-wide assessment that is similar to genome-wide assessment. The exposome needs scale and speed and this may not require the same level of precision that the field of analytical chemistry needs. Sacrificing some sensitivity to increase throughput and reduce cost could enable the testing of more than 100,000 samples within the confines of existing research budgets. The technologists must understand this problem and that there is a potential market for such instrumentation. Current high-resolution mass spectrometers can have long run times that restrict throughput to 40 to 60 samples per week; thus, finding innovative solutions to reducing the run times would improve the field.

The aptly named Agilent Rapid Fire 365 Mass Spectrometer replaces the slower liquid chromatography separation with an automated solid-phase extraction. This can be coupled with triple quadrapole or time of flight mass spectrometers to increase throughput 1,000-fold. It is possible to run well over 10,000 samples per week. Although there is a concomitant loss in separation and resolution, there is a dramatic increase in throughput. A similar system is not available for the Orbitrap instruments, but from a technical standpoint there is no reason that the basis of these technologies cannot be combined to create a system that could increase throughput by 10 times while reducing individual sample cost by 75 percent. A team of investigators and engineers could devise the optimal balance among sensitivity, resolution, and throughput that could put these advanced systems into the framework needed for almost any large-scale human study.

Of course, the amount of data generated could increase by 20 times for the same research dollar spent, indicating a need for a similar level of innovation on the bioinformatic side. Automated bioinformatic pipelines would need to be developed in parallel. On the bioinformatic side, I believe that once we are able to conduct an exposome-wide study on 200,000 people, we will then have enough data to create a reference map of the exposome. A rough reference map could be achieved with 10,000 to 20,000 subjects. With data from such a large number of subjects that includes information on over 100,000 chemical features, we will be able to determine which measures are robust and reliable, which chemicals are found in a representative population, and which chemicals are key drivers of health and disease. From there, data dimensionality reduction strategies developed for this specific purpose can likely decrease the bioinformatic and statistical burden from future studies. But we first need the large-scale data to guide these efforts.

The machine learning and artificial intelligence approaches described in Chapter 8, Data Science and the Exposome, could ultimately lead to an automated pipeline that identifies and annotates features, queries databases for matches, clusters common exposures, compares exposures to disruptions in metabolic pathways, creates causal maps between exposures and health outcomes, compares the exposure–response relationships to those causal maps, and then presents the data in a form suitable for publication. This fanciful workflow may seem impossible now, but each and every one of these steps has already been automated for some purpose, whether it be in chemistry, genetics, computer science, statistics, or the financial sector. My potentially naïve stance is that our current limitations are self-imposed. With the proper priorities and resources, these innovative solutions could be manifested within years.

6.8 The science of innovation

Science, the Endless Frontier (Bush, 1945) is cited as the blueprint for the need for scientific advancement in society and provides a framing for innovation (see Fig. 6.2). Vannevar Bush led the Office of Scientific Research and Development, a unit that was essentially created for him by U.S. President Franklin D. Roosevelt to propel American science after World War II. I cited Bush's work in an editorial where I wrote about the importance of science in society (Miller, 2017). In Bush's words:

> *The pioneer spirit is still vigorous within this Nation. Science offers a largely unexplored hinterland for the pioneer who has the tools for his task. The rewards of such exploration both for the Nation and the individual are great. Scientific progress is one essential key to our security as a nation, to our better health, to more jobs, to a higher standard of living, and to our cultural progress.*

Of course, this was written in the context of the United States, but this spirit can be found throughout the world, and scientific progress is key to securing the security of our planet. Another major edit I would make in his quote is the phrase "*his* task." Fortunately, the pioneers of

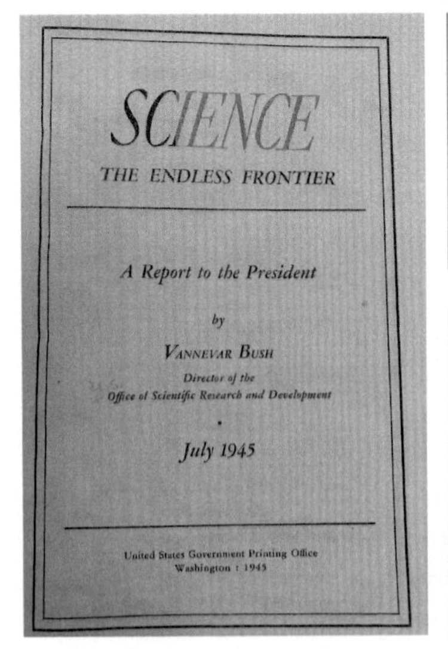
SCIENCE
THE ENDLESS FRONTIER

A Report to the President
by
Vannevar Bush
Director of the
Office of Scientific Research and Development

July 1945

United States Government Printing Office
Washington : 1945

SCIENCE—THE ENDLESS FRONTIER

"New frontiers of the mind are before us, and if they are pioneered with the same vision, boldness, and drive with which we have waged this war we can create a fuller and more fruitful employment and a fuller and more fruitful life."—

Franklin D. Roosevelt.
November 17, 1944.

LETTER OF TRANSMITTAL

Office of Scientific Research and Development
1530 P Street, NW.
Washington 25, D. C.

July 5, 1945.

Dear Mr. President:

In a letter dated November 17, 1944, President Roosevelt requested my recommendations on the following points:

(1) What can be done, consistent with military security, and with the prior approval of the military authorities, to make known to the world as soon as possible the contributions which have been made during our war effort to scientific knowledge?

(2) With particular reference to the war of science against disease, what can be done now to organize a program for continuing in the future the work which has been done in medicine and related sciences?

Figure 6.2 ***Science: The Endless Frontier.*** This report provided a vision for science from Vannevar Bush to President Franklin D. Roosevelt. It was published in 1945 for consumption by the general public. It was received with a high degree of enthusiasm. Even though it was written 75 years ago, its principles provide a defense of science for the sake of science, and all current scientists are encouraged to read it.

today are just as often women as they are men. The historic exclusion or discouragement of women in science has deprived us of half of our intellectual power, and as a society we would be wise to ensure that all innovators needed to advance science are given the support necessary for success (for an example, see Moss-Racusin, Dovidio, Brescoll, Graham, & Handelsman, 2012).

As detailed in Chapter 1, The Exposome: Purpose, Definitions, and Scope, we have made outstanding progress in understanding disease pathogenesis, but the majority of the effort has focused on the genetic drivers of disease. We have been ignoring our environment as the key driver it is. Our current system is not providing the right incentives for innovation on the environmental side. I opened this book with a quote of Edwin Land of Polaroid fame. "Don't undertake a project that is not seemingly impossible and manifestly important." Is this what the exposome is trying to do? Is it manifestly important? As described in the opening chapter, our phenotype results from a combination of our genes and environment. The tools to assess the environmental contributors are not as robust or comprehensive as those for the genetic contributors. Thus, if we care about what makes us human, what promotes longevity, and what causes disease, then we better invest in getting the right information. Manifestly important, indeed! Peter Medawar captured the essence of what happens when intellectually creativity is combined with an important question. He was making the point that there were many exceptionally intelligent scholars at Oxford and Cambridge, but that James Watson's scientific achievements were also propelled by other factors. Medawar notes that Watson had a certain level of luck when he stated, "Watson had one towering advantage over all of them; in addition to being extremely clever he had something important to be clever about." (Medawar, 1984). The DNA revolution— from the elucidation of the structure of DNA to the Human Genome Project, to genome-wide association studies, CRISPR, and gene therapy—benefitted from the "something to be clever about." If we believe the truism that who we are is a combination of our genes and environment, should not the exposome be "something to be clever about?"

I also appreciate Bahcall's physicist-based take on innovation and his similar questioning the limits of the term "moonshot." *Loonshots* provides a rhyming alternative with a subtle insertion of the crazy (the book's subtitle is "How to nurture the crazy ideas that win wars, cure diseases, and transform industries"). My mind goes to the aviary variety of loon, as well as the Looney Tunes cartoons that shaped my childhood. If one looks up the term "loon," one will see many photos of birds and references to a silly or foolish person. Even with my own misgivings of the term "moonshot," loonshot may also be slightly off the mark. Bahcall's work not only is insightful and filled with intriguing examples of innovation, but also appears to be one of those wonderfully compilated set of anecdotes that leaves one wondering about cause and effect. I refer back to the work of Rosenzweig in *The Halo Effect*, which essentially designed a randomized control trial of corporate success. I would welcome a similarly astute examination of innovation as it relates to scientific achievement.

That said, Bahcall's conclusions do appear to be supported by a wide range of evidence. One of the key take-home messages is the need to coordinate the support of the generation of crazy ideas with the systematic franchise-type work that puts those ideas into practice. For a scientist, I believe this is analogous to the combination of high-risk and low-risk projects. I routinely have students working on both. The low-risk, franchise-like projects follow up on key ideas in the lab. They build upon years of similar projects and create output similar to the lab's franchise.

They teach the student how to formulate hypotheses, generate high-quality data, interpret that data, and write manuscripts. At the same time, I encourage students to consider the types of projects that are daring and that could disrupt the lab's franchise work. In *Loonshots* Bahcall emphasizes the need to have communication between the teams. When a person performs both types of projects, there is a need for internal compartmentalization, but it is much easier to open up the lines of communication when the voices are all inside the same head. Within small teams, it is also possible for some people to be responsible for the crazy ideas for one project and the franchise work of another project. Indeed, a properly conducted lab meeting should foster this type of collaboration. It will be interesting to see if some of the new projects on the human exposome can achieve this balance or if they will be overly weighted to franchise-level work.

This systematic juxtaposition between the creative (thinkers) team members and those who conduct the meticulous studies that generate the products and services that are necessary (doers) is easier said than done. Often it is the lab director that plays the role of thinker with the minions of trainees being the doers, but this does not lead to innovation; it merely advances the franchise of the lab director. A key ingredient for success in any scientific organization is that transparent recognition that both components are necessary. Too often the creative Steve Jobs types get all of the credit when in fact the great ideas would never be transferred into reality without the similarly brilliant, but quietly audacious Steve Wozniak types.

This is why I have been especially excited by the efforts to evaluate "the science of science" as exemplified by Laszlo Barabási's group. In the wonderfully repetitive "The Science of Science" published in *Science* (Fortunato et al, 2018), the authors use advanced data analytics and a trove of available publication and citation data to examine the features of innovation and breakthrough research. It is a quantitative and not anecdotal approach. It overturns long-term myths about younger scientist being more creative and inventive and examines optimal science team sizes. For the purposes of the exposome, the paper demonstrates the essential nature of interdisciplinary collaboration and a balance of fundamental and risky efforts for major breakthroughs. Research groups working on the exposome must pursue technologically innovative projects, while at the same time, conducting studies that build upon our existing knowledge base. As I describe in Chapter 10, The Exposome in the Future some of the early exposome projects focused on the latter and did not invest sufficiently in the former. I will admit that the HERCULES Exposome Research Center at Emory was similarly imbalanced, but in the opposite direction. We focused on innovation and did not worry much about doing the normal science that is critical to convince the field that what one is doing is practical. Those demonstration projects took the form of separate research projects on particular diseases or conditions. The idea was that the center would spawn offshoot projects (the NIH R01–type grant) that built upon the innovative techniques, but focused on a specific disease process or condition. In this scenario the innovation hub bears no direct responsibility for the work being conducted allowing the innovation team to continue to innovate without having to worry about specific deliverables. Not that there was not accountability, but rather the deliverables were more about enabling other people to do things they were unable to do previously. Progress was measured by evidence that the new tools were being developed, and less so on them being incorporated into research projects. That is much different than quantifying the number of samples run or the number of papers published. When scientists are judged primarily on the volume of their output, there is a strong tendency for their innovation to be muted.

In the book *Innovation: A Very Short Introduction*, the authors Dodgson and Gann cite the innovative efforts of Josiah Wedgewood as he built the hugely successful Wedgewood China industry. The authors detail the various efforts of Wedgewood to systematize the design and manufacturing of his products, but not necessarily reduce costs. Indeed, his products were valued for their quality and he was able to charge *more* than his competitors. Wedgewood is praised for being an innovative and business-savvy artisan. What does fine China have to do with science? In a genetic link back to our history in the first chapters, Wedgewood's grandson was none other than Charles Darwin. Were the innovative, industrial, and organizational traits of his grandfather passed down to Darwin? Or was it the familial financial resources afforded Darwin that allowed him to take his fantastic scientific voyage? Perhaps both, but regardless, it serves as a reminder that scientists need to be looking to other sectors to learn how ideas are developed, systems are optimized, and new solutions are brought to bear, as well as being cognizant of the market conditions and business decisions that support innovation. The ivory tower may be a good place to contemplate, but it is insufficient place to innovate at the scale needed for the exposome.

6.9 Collaborative research

There are several lines of evidence that major scientific advances occur when different groups or disciplines are brought together to solve a problem (see Fortunato et al., 2018). This is why administrators value and promote interdisciplinary, collaborative, and team science. The value of collaborative research is illustrated by the late Peter Medawar:

> *The rationale of collaborative research is the synergism of two or more minds working towards the solution of the same problem (two or more people working together can accomplish more than the sum of what would have been possible if those same people had been working on their own)*
>
> ***Medawar (1988)***

He goes on to note that science and technology are probably the best areas for this. For example, it is unlikely that a novel would be better if it were written by two authors or a painting or sculpture created by three for four masters. I would argue that the broader the expertise and knowledge of the collaborative team, the greater the potential for novel and innovative solutions. This is why interdisciplinary research has such value. The nature of collaboration is implicit. What Wild pointed out when he coined the term "exposome" was that the epidemiologists were not able to embed all of the exposure assessors' measures into their epidemiological framework the way that genetics had been able to do. Everybody knew that the environmental factors were important, but how could one link the external forces with biological alterations and health outcomes?

One of the most critical components of innovation is the recruitment of individuals who possess knowledge in different fields and domains to apply their way of thinking to the new problem. Convergence among different fields is repeatedly cited as a key ingredient for innovation. I must admit that this has been my primary approach to advancing the exposome. I have been systematically recruiting investigators to the field with diverse backgrounds to increase the quantity and quality of my own competition. I want the greatest minds to help define and measure the exposome.

Bahcall discusses the limitations of "cultures of innovation." People often harp about the importance of culture, but this is one of the traps often seen along the lines of business books like *Good to Great*, which use superficial analyses. By culture in an organization setting, I mean the values and beliefs that drive the interactions among the workers and with customers. This should remind the reader of Kuhn's paradigm definition described in Chapter 1, The Exposome: Purpose, Definitions, and Scope i.e. the entire constellation of beliefs, values, and techniques shared by the members of a community. Incredibly innovative organizations often have great culture because the work is fun and rewarding. Thus the innovation is driven by the nature of the work rather than the culture of the organization. Just like in protein chemistry where function follows structure, so does culture. A culture conducive to innovation requires *structures* that incentivize the appropriate behavior. Well-designed organizations with proper management support and structure increase the probability of success and can, in turn, enhance culture.

Bahcall discusses his formula for loonshot success. The first part is separating loonshot and franchise groups in what he refers to as phase separation. The second is creating a system of seamless exchange between the two groups in what he calls dynamic equilibrium. The last is the development of a critical team mass that allows new ideas to ignite and take off. From a corporate standpoint, one can consider the business idea of making a strong flavorful coffee as the loonshot and the franchise as the disciplined business process development needed to make the idea successful, much like the franchising of a coffee shop or restaurant. Although the business world is full of examples where one or two people conceived an innovative idea and built the foundation for the company without all three pieces of this planning system (phase separation, dynamic equilibrium, and critical team mass). However, if you are an organization charged with creating innovation, there is pressure to at least have some sort of plan. For the sake of illustration, let us see if we can put a hypothetical exposome innovation team together (Fig. 6.3).

Here I describe the three components of the innovative teams that exposome research will need.

1. Creators. The creators focus on technological development and improving measurement, which includes increased sensitivity, greater resolution, better identification, higher throughput, and lower cost. This technological tinkering should be pursued without regard to how it will be used. The team should focus on making better instruments. At the same time, the bioinformatic and data analytics platforms must be improved. How do we interpret multiomic information? What can we learn from genomics and proteomics? Although our exposome data are more complicated than high-throughput genetic sequencing and the algorithms that were developed to piece together the long

Innovation steps for the human exposome

(1) Technological advances: sample collection, high-resolution mass spectrometry, sensors, robotics, automated bioinformatic and data science pipelines, improved mass spectral libraries for identification

(2) Well-designed human studies: representative cohorts with multiscale environmental measures and biobanked samples amenable to advanced technologies

(3) State-of-the-art technologies combined with state-of-the-art cohorts (potential merged/virtual)

Figure 6.3 ***Steps toward exposome innovation.*** Based on the principles outlined by Bachall in *Loonshots*, this figure demonstrates how separating the different subdisciplines can help an organization achieve innovation. No. 1 represents the technological innovation that may at times seem fanciful and unachievable. No. 2 represents the normal science of studying human disease. No. 3 represents the convergence of advances in nos. 1 and 2. It is critical that as the first two stages are pursued that there are frequent opportunities for the exchange of ideas between the two teams, but that the majority of the work occurs separately.

reads from the DNA analyzers, the genomic successes do illustrate the combined need of analytical chemistry and bioinformatics. Will there need to be dimensionality reduction to fit the HRMS square peg into the epidemiological round hole?

2. Implementors. These are the franchise correlates. It is the integration of the information derived from the creators in no. 1 to studies of human health. This is where study design comes in. For the exposome one could consider the classically trained epidemiologist with a grounding in Bradford-Hill criteria as the franchise-level work. Well-characterized human populations with serial biological and environmental samples being collected over time will provide the scientific customer base or need. It is quite possible that many of the human studies currently underway can provide the exposure data and the biological samples that can be analyzed with the technologies developed under no. 1.
3. The dynamic equilibrium combined with critical mass. This is what I view as the true human exposome project. It is when we can design a series of studies with the appropriate cohorts and the innovative tools to draw causal links between complex exposures and complex diseases. Could this come out of the Precision Medicine Initiative's program All of Us in the U.S.? Could it result from a merging of cohorts being funded by the European Commission's Human Exposome Project

Network? The critical mass will result from having an international team of researchers willing to share data, ideas, and resources to conduct the types of studies that will provide unbiased and systems level analyses of the exposome. At this time, I do not see the appropriate intellectual interface among the scientific team members (the dynamic interface). That is, there is still a major chasm between the chemists/ biologists and the epidemiologists. If we want to make associations between exposures and disease, we must have rigorous study designs, but at the same time, these procedures must develop strategies to incorporate the massive amount of information coming from HRMS and other measures of external exposures and biological responses.

At this point in time, a comprehensive analysis of chemicals in plasma, either as a list of thousands of features or as networked exposures or biological responses, does not fit into the current way epidemiologist do their work. This is where the team separation is important. If the research group developing innovative measurements is being guided by managers who do not see how the integration could ever happen, it is highly unlikely that the innovative work will be supported. Exposome research teams must enhance the intellectual equilibrium between those working at the molecular level and those working at the population level. What can the exposome learn from other sectors? There are several outstanding examples of innovative organizations, but their successes are far from formulaic. Bell Labs, IBM, Apple, Microsoft, and Roche Institute of Molecular Biology all recruited extraordinarily talented individuals and they were given intellectual freedom, but it was clear for whom they were working. Recruitment of the individuals who have the ability to innovate and then allowing them to do that is one of the most important components of success. That said, it can be difficult to colocate multiple brilliant scientists without proper management and leadership.

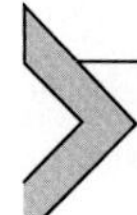

6.10 Providing evidence for policy and behavioral change

The increased demand for solutions to our changing climate has primarily focused on curtailing certain activities—for example, reducing fossil fuel use, reducing waste, altering eating patterns, and reducing consumerism. All of these approaches are admirable and should be pursued.

However, the link to global temperatures is often indirect and separated by time and space, such that it is extremely difficult to demonstrate to the public the impact of their efforts on the long-term goal of reducing atmospheric carbon dioxide and global temperatures (one must rely on rather sophisticated mathematical models that the general public do not comprehend). However, the effects of shifting diets, reducing use of gas-powered vehicles, and eliminating coal-fired power plants have an almost *immediate* effect on people's health in their local environment. Changes that can be evaluated in near real time using the HRMS approaches described earlier and in Chapter 5, Measuring Exposures and their Impacts: Practical and Analytical, provide an opportunity to influence behavior by providing the right type of data in the right setting in a timely fashion. Policymakers and citizens need data demonstrating the positive impact from their modified behaviors, and this must occur in the short term. If a person is able to see that the purchase of an electric or hybrid car reduces the amount of vehicle emissions *in their own body* within a week of purchasing the car, that could be more influential than the savings at the gas pump. If the mayor of a city sees reduced hospital costs for asthma-related conditions within months of installing traffic control measures in congested parts of the city, it will reinforce and amplify these efforts. Innovation is not all about technology. How we engage the general public, how we implement policy, and how we evaluate the impact of interventions all have benefited from creative new approaches. In a changing world, there is a continual need to improve our systems to allow adaptation to external forces.

6.11 Obstacles and opportunities

The opportunities for innovation in exposome research are endless. The exposome presents major technological and bioinformatic challenges. Given the proverbial driver of invention being necessity, once the biomedical, statistical, and data science communities embrace the exposome as a necessity, there are numerous opportunities for innovation. The scientific community recognized the importance of genomics early on and there was broad support for research in this area, even though there was some disagreement on whether a Human Genome Project was the best way to allocate resources. As will be outlined in the closing chapter, it is the articulation of an ambitious vision for exposome research with a set of milestones that will be necessary for the scientific community to see it as a valuable target. Scientists and engineers will need to see that the field is moving in a direction in which inventions and efforts will be rewarded.

As noted in the discussion about paradigms in Chapter 1, The Exposome: Purpose, Definitions, and Scope, major shifts within fields are often accompanied by a significant amount of discomfort. Scientists spend in excess of 10 years learning to be a scientist. When confronted with approaches that they did not learn in school, many worry that their training was insufficient for the new challenges and that they must learn more tools and techniques. Innovation can conjure up similar levels of discomfort and concern. Fortunately, if organizations make their approach to innovation to their employees clear (well-articulated goals, ample opportunities for training and retooling, and reward systems that support innovative efforts), they can avoid some of the angst and increase the likelihood of remaining relevant or even in business. Academic programs are also under pressure to innovate or fall to those fields and disciplines that are constantly updating approaches and refusing to persist with the status quo.

6.12 Discussion questions

1. Imagine that you are charged with establishing an innovation team at your organization (educational, research, corporate). What are the types of people and systems you would put in place to provide sustained innovation?
2. What is the most innovative situation in which you have participated? What were the features or components that supported innovation?
3. What do you see as current bottlenecks to exposomics and which of these problems would you improve with innovative ideas first?

References

Bush, V. (1945). *Science: The endless frontier*. Washington, DC: United States Government Printing Office.

Fortunato, S., Bergstrom, C. T., Borner, K., Evans, J. A., Helbing, D., Milojevic, S., … Barabasi, A.-L. (2018). Science of science. *Science*, *359*, eaao0185.

Medawar, P. (1984). *Pluto's Republic: Incorporating the art of the soluble and induction and intuition in scientific thought*. Oxford: Oxford University Press.
Medawar, P. (1988). *Memoir of a thinking radish*. Oxford: Oxford University Press.
Miller, G. W. (2017). Science, society, and societies. *Toxicological Sciences, 156*(1), 2–3.
Moss-Racusin, C. A., Dovidio, J. F., Brescoll, V. L., Graham, M. J., & Handelsman, J. (2012). Science faculty's subtle gender biases favor male students. *Proceedings of the National Academy of Sciences of the United States of America, 109*(41), 16474–16479.
Pisano, G.P. (2019). The hard truth about innovative cultures. *Harvard Business Review, (January-February).*
Shreeve, J. (2004). *The genome war: How Craig Venter tried to capture the code of life and save the world/James Shreeve*. New York: Alfred A. Knopf.

Further reading

Bahcall, S. (2019). *Loonshots: How to nurture the crazy ideas that win wars, cure diseases, and transform industries*. New York: St. Martin's Press.
Collins, J. (2001). *Good to great: Why some companies make the leap...and others don't*. New York: Random House Business Books.
Dodgson, M., & Gann, D. (2018). *Innovation: A very short introduction* (2nd ed.). Oxford: Oxford University Press.
Eliuk, S., & Makarov, A. (2015). Evolution of orbitrap mass spectrometry instrumentation. *Annual Reviews of Analytical Chemistry, 8*, 61–80.
Griffiths, J. (2008). A brief history of mass spectrometry. *Analytical Chemistry, 80*(15), 5678–5683.
Rosenzweig, P. (2007). *The halo effect: And the eight other business delusions that deceive managers*. New York: Free Press.

CHAPTER SEVEN

Pathways and networks

The clearest way into the Universe is through a forest wilderness.

John Muir

7.1 Introduction

Although a quote from a naturalist and preservationist is certainly appropriate in a book about the exposome and the importance of the environment in health, I am using it here in a different context. Muir is making the point that the most direct way to understand the complexity of the universe is to enter the dense forest of the unknown-to explore areas where it can be difficult to see one's path. Being lost in nature may well be the best means of finding our way. When one walks through the woods, whether it be giants like Muir, Carson, or Thoreau, or graduate students taking needed time away from the laboratory, the repetitive nature of trees, plants, and paths allows one to contemplate and reflect on one's position the world. Ironically, the simplicity of nature is anything but simple. The shape of individual leaves, the organization of the tree itself, and the rocky outcroppings are beautifully mathematical, containing ubiquitous fractal patterns à la Benoit Mandelbrot. The organization of our world and universe can be described simultaneously with simple analogies and complex mathematics. This chapter describes approaches that allow us to distill the complexity of our environment, but do so in a way that is amenable to mathematical solutions. I focus on the use of pathways and networks to better understand the complex nature of the exposome and suggest how these tools can help make the exposome soluble.

Pathways force scientists to step back from their reductionism and examine diverse downstream effects. Metabolic pathways, adverse outcome pathways (AOPs), and aggregate exposure pathways (AEPs) will be integral to the field (although pathways may not be the most appropriate description, as most of these concepts are actually exhibited as networks).

The Exposome.
DOI: https://doi.org/10.1016/B978-0-12-814079-6.00007-9

Advances in network science are providing a way of organizing exposome-related data in a way that will improve the conceptualization, visualization, and understanding of the complexities of exposures and responses. This chapter reviews the pathway- and network-based approaches that will support the future of exposome research.

7.2 Biological pathways

Simple chemical reactions illustrate how two substrates can combine to make a product. When there is a catalyst present, the process accelerates. When these chemical reactions move into biological systems, the processes are not that different. Biochemistry focuses on the pathways that move substrates to products. I learned about numerous biochemical pathways throughout my education, with enzymes, cofactors, and feedback inhibition loopsfilling my head for many years. Michaelis–Menten equations helped me understand the kinetics, but I did not spend much time examining the reaction rates in a quantitative or mathematical manner. After learning more about brain neurochemistry and moving into a faculty position, I started thinking more about the quantitative aspects of the neurochemicals I was studying in the laboratory. A fortuitous meeting over 15 years ago with a faculty member at Georgia Tech, mathematician and systems biologist Eberhard Voit, helped expand my thinking.

Eberhard and I applied for some internal university funding to apply his expertise in systems biology to the Parkinson's disease-related dopamine neurochemistry I was studying. We received the funding, hired a postdoc (Zhen Qi), and started our collaboration. I knew neuroscience and toxicology, and Eberhard knew mathematics and systems biology. There was not very much domain expertise overlap. We would sit for hours as he asked me questions about enzyme reactions, toxic intermediates, and key components. I distinctly recall him asking me questions that I could not possibly know and then saying "Just give me your best guess, would you expect more or less of that reaction product?" I could not understand how, at that moment, precision did not matter. We went through this exercise for dozens of reactions, combed the literature, and built the biochemical model. Eberhard had been a leader in the area of biochemical systems theory (BST), which allows for modeling and prediction of biochemical pathways under a minimal set of assumptions and is even forgiving when crucial quantitative information is unavailable. BST is based on the idea that the reactions are modeled as products of power-law functions. The representation should be appropriate in the vicinity of a particular biological setting, and it is acceptable for this setting to be complex, as such complex systems are well characterized by power-law functions. The resulting equations can model any possible nonlinearity that has continuous derivatives, including limiting cycles and deterministic chaos. Proteins and metabolites are considered dependent variables in the systems, while constant inputs or enzyme kinetics that do not change in a given experiment are considered to be independent. Even though many of the actual values are unknown, the structure of the equations is known and the unknowns can be represented symbolically. Fig. 7.1 shows the initial dopamine metabolic pathway we developed (Qi, Miller, & Voit, 2008a). Each reaction is represented with a differential equation. We then performed a series of computational simulations on the entire system to see how downstream metabolites would change based on alterations to particular enzymes or substrates. Here we had some useful experimental data from a range of transgenic mice in which enzymes or transporters had been genetically removed (either 100% for a knockout or 50% for a heterozygote knockout). The results were striking and the predictions aligned very well with the neurochemical analysis. This initial foray into systems biology helped me appreciate the utility of these approaches to build predictive models and quantify these important processes. Later, we examined how a range of environmental chemicals perturbed dopamine neurochemistry using the models we had constructed (Qi et al., 2008a; Qi, Miller, & Voit, 2008b; Qi, Miller, & Voit, 2009; Qi, Miller, & Voit, 2010; Qi, Miller, & Voit, 2014; Voit, Qi, & Miller, 2008). Later as we developed the HERCULES Exposome Research Center, I recruited Eberhard to lead a core in systems biology to assist us as we grappled with the forthcoming complexity.

Before my collaboration with Eberhard, my strongest memory of biological pathways was the dauntingly complex biochemical maps from Sigma Chemicals that graced the walls of many laboratories. Graduate students learned biochemistry as one isolated reaction at a time and then particular multistep pathways such as Kreb's cycle. One could intuit that each reaction would likely be affected by the other reactions and pathways occurring within a cell, but systems biology was not taught at that time so we continued along a rather reductionistic route. The Kyoto Encyclopedia of Genes and Genomes (KEGG) was launched in 1995 to

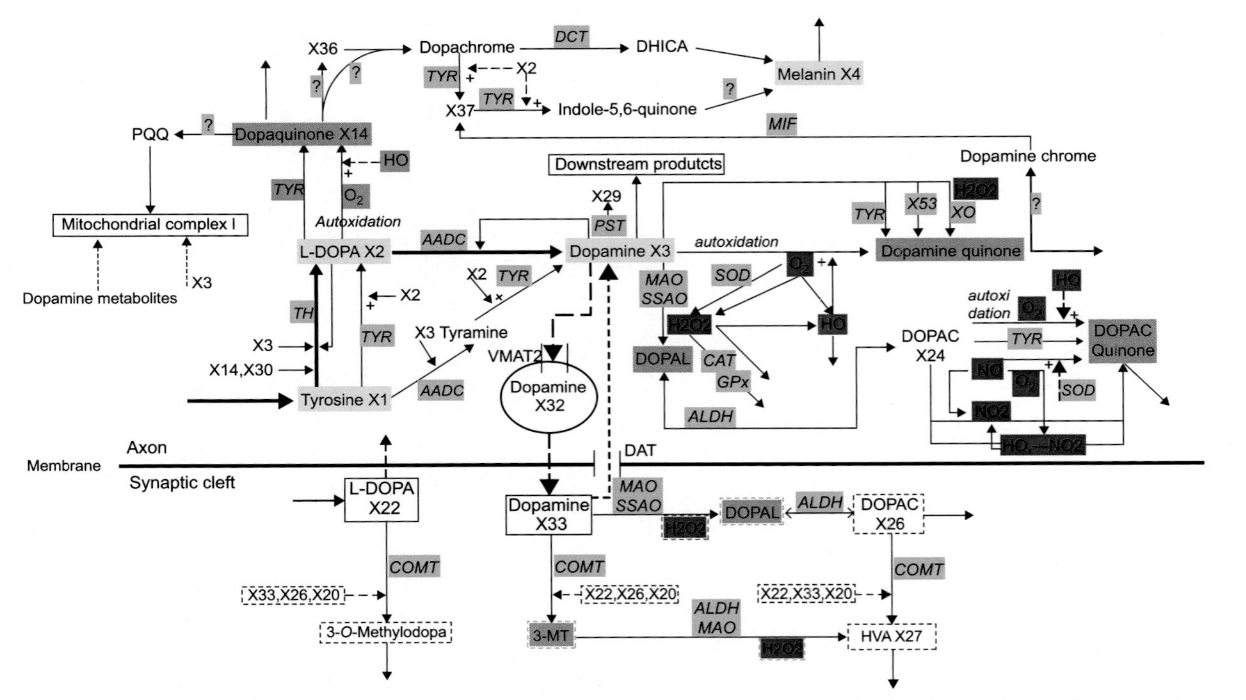

Figure 7.1 ***Simplified diagram of the dopamine biochemistry.*** Primary metabolites are highlighted in yellow (lightest shading), reactive oxygen and nitrogen species in light red (darkest shading), and toxic species in light purple (medium shading). The ellipse shows dopamine inside vesicles. Metabolites in the synaptic cleft are indicated by dashed frames. Solid arrows represent biochemical reactions; associated enzymes are denoted in capital italics in light blue (medium shading with italics). Dash-dotted arrows designate inhibition, while dashed arrows with plus sign designate activation. Transport steps are represented as dashed arrows. Question marks refer to enzymes that are unknown or not fully understood. *AADC*, DOPA decarboxylase; *ALDH*, aldehyde dehydrogenase; *CAT*, catalase; *COMT*, catechol *O*-methyltransferase; *DCT*, dopachrome isomerase; *GPx*, glutathione peroxidase; *MAO*, monoamine oxidase; *MIF*, migration inhibitory factor; *SOD*, superoxide dismutase; *SSAO*, semicarbazide-sensitive amine oxidase; *TH*, tyrosine hydroxylase; *TYR*, tyrosinase; *XO*, xanthine oxidase. *Reproduced with permission from Qi, Z., Miller, G. W., & Voit, E. O. (2008a). Computational systems analysis of dopamine metabolism. PLoS One, 3(6), e2444.*

provide a map of the genetic pathways that were being discovered during the Japanese Human Genome Project (Kanehisa, Goto, Sato, Furumichi, & Tanabe, 2012; Ogata et al., 1999). By showing the wide range of pathways, and their interconnectedness, it provided scientists a means of integrating the complex data that were being generated. It included more than the genes—it included the encoded gene products, such as RNA, proteins, and metabolites. KEGG modules provided increasingly rich information about particular pathways. Drugs and their active metabolites have been also been incorporated, as well as known genetic and environmental contributors to many diseases. Unfortunately, even if we know from epidemiology that a certain environmental chemical contributes to a disease, we generally do not know with any precision what the specific biochemical perturbations are that cause the disease, leading to many gaps in knowledge. Even so, for most human metabolism the KEGG pathways have been an organizing framework that has helped investigators interpret data from metabolic-based research. Fig. 7.2 shows an overall metabolic map from KEGG. It is difficult to discern any specific pathway from the

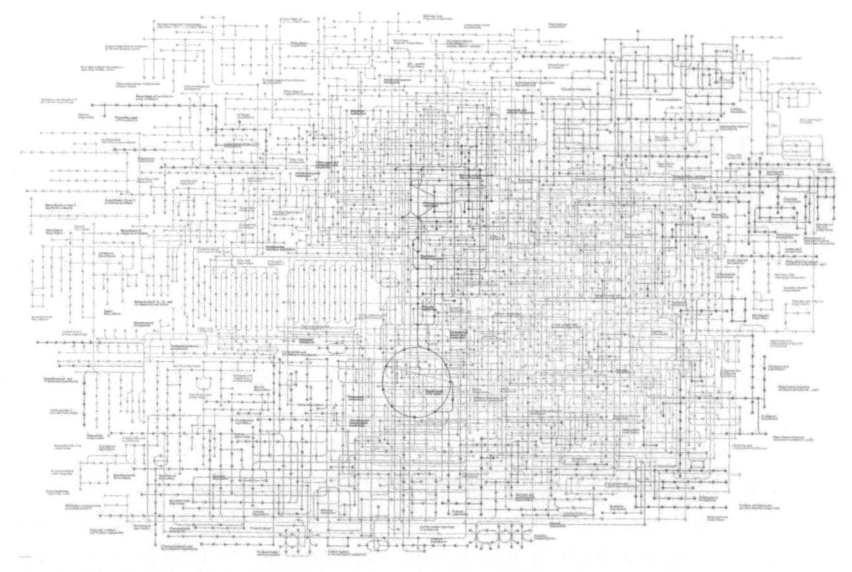

Figure 7.2 ***Large-scale metabolic map.*** This metabolic map provides a visualization of the major metabolic pathways in humans. The sheer complexity of the thousands of biological pathways illustrates the complexity of human biochemistry. The difficulty in evaluating how complex environmental exposures disrupt systems-level metabolism is apparent from this complex system. *The image is reproduced from KEGG with permission.*

global map, but it helps illustrate the forest we must explore in order to understand our biochemical universe. There are numerous biological reactions occurring with our cells and tissues, and they may be working in concert or in competition. Adjacent cells may have different components and pathways. Different organs and tissues have specific pathways for their particular functions. Thus, each cell is a microcosm of human metabolism. When one contemplates the breadth of cellular diversity, it can be haltingly complex; however, at the same time, when one considers how much information can potentially be gleaned from a drop of blood, some of the anxiety over the complexity subsides. When we see biomarkers of a heart attack in a blood sample, it tells us that one particular organ can release a metabolite that becomes systemic and measurable. The release of insulin from the pancreas sets off a cascade of systemic changes. Minute amounts of hormones from the tiny pituitary gland can cause dramatic physiological functions. All of clinical chemistry is based on the concept that our circulating blood represents that soup of biology in which we live. Peripherally sampled blood, while not perfect, does provide an extraordinary multifactor integrating biological source of information.

7.3 Exploring the pathways

The simple fact that our blood carries such important information collected from throughout the body is why the field of clinical metabolomics has expanded so rapidly. There is so much to be learned. I first started working with Dean Jones and untargeted metabolomics nearly 15 years ago. As he tried to make sense of the vast amounts of data, he explained work that an Emory colleague was doing on pathway analysis. Shuzhao Li was a young investigator working on metabolomics in collaboration with several groups. His past work on the mummichog fish (*Fundulus heteroclitus*) led him to name his new approach after the fish. He introduced the pathway analysis tool mummichog as an alternative to the classical way of conducting untargeted metabolomics. The conventional method requires that the metabolites be rigorously identified prior to conducting pathway or network analysis. Identified metabolites could then be mapped into the known KEGG pathway maps. Mummichog took a different approach. The algorithm predicted functional activity with minimal knowledge on the identification of the metabolite. For example, one may

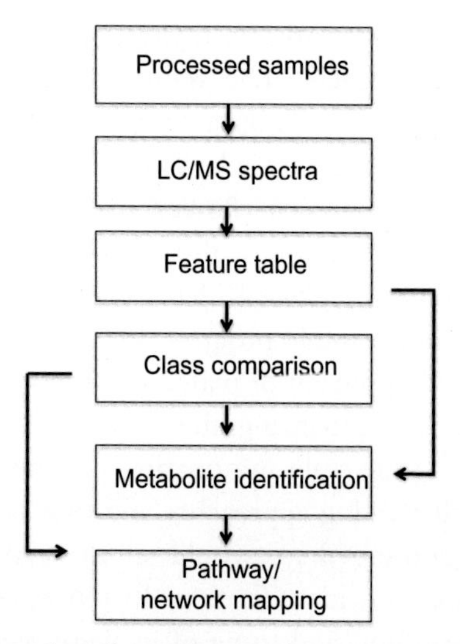

Figure 7.3 ***Mummichog-based analysis of mass spectrometry data.*** This figure illustrates the key steps used during the pathway-based mummichog program. Rather than waiting for precise chemical identification, mummichog uses approximations from *m/z* and retention values to predict possible matches and uses these estimates to map the features. Random features are randomly distributed across biological pathways, whereas nonrandom features will cluster within affected biological pathways.

have the mass/charge (*m/z*) value that could be 1 of 10 different metabolites. The odds of correct placement are relatively low. But mummichog maps *all* of the *possible* metabolites much like if every color of paint was randomly splattered against a wall (Fig. 7.3). If a chemical exposure or disease state perturbed Kreb's cycle and all of the metabolites splattered on the metabolic wall were false metabolites or false positives (those that are unassociated with an exposure or a disease state), they would be distributed randomly. However, those that are true metabolic matches will exhibit a local enrichment at the target site—in this example, Kreb's cycle. Thus, mummichog is using all of the available information and simply distinguishing between random and nonrandom associations. The initial challenge was that the classical analytical chemist would be apoplectic with using poorly identified features to identify metabolic alterations. Indeed, this approach was met with considerable resistance for several years. Fortunately, The Jones research group and my own laboratory did not merely trust the computational approach but used orthogonal measures to verify the accuracy of the predictions. This included targeted biochemical assays, protein chemistry, replicate experiments in animal and cell models, and many others. The results were simply extraordinary. Mummichog could provide biological insight with relatively low-level analytical confidence in the features identified from the mass spectrometry. My trust in this type of approach allowed me (and the Emory team) to have a reasonable level of comfort with the uncertainty of the mass spectral peaks. Yes, we wanted high-confidence identifications when possible, but when they were not possible progress was not halted. In my opinion, the mummichog-based approach provided the Emory team with the type of saltatory leap in conceptualizing the exposome. We knew that we would be able to get omic-scale information on endogenous and exogenous metabolites with existing approaches and that as the technology and informatic advances continued we would see the exposome biochemistry with greater resolution and identification. In retrospect, this work was at the early stages of loonshot, and some of us were recognizing that it was slightly less crazy than it had first appeared. Jeff Xia, who developed Metaboloanalyst (metaboanalyst.ca/faces/home.xhtml) and was a university classmate of Shuzhao Li's, has incorporated mummichog into his computational website. Other groups are also adopting this innovative approach to their metabolomics workflows.

7.4 Adverse outcome pathways

Regulators across the world have been grappling with providing useful information to inform policy. The one chemical at a time approach has not made a dent in the more than 80,000 chemicals for which we would like information. It was clear that many chemicals were converging on the similar biological pathways. For example, Fig. 7.4 shows the KEGG map for chemical carcinogenesis for one class of chemicals. This type of information indicates that we have substantial knowledge of how some chemicals exert their adverse effects. This led many teams to consider the idea of creating pathways that describe the adverse consequences. This would not merely be a KEGG-based metabolic map, but rather a system that would use toxicological and mechanistic data that could describe the molecular targets and key events (KE) that led to toxicity.

There has been global recognition of the utility of the adverse outcome pathway (AOP) concept. One of the most important steps was the

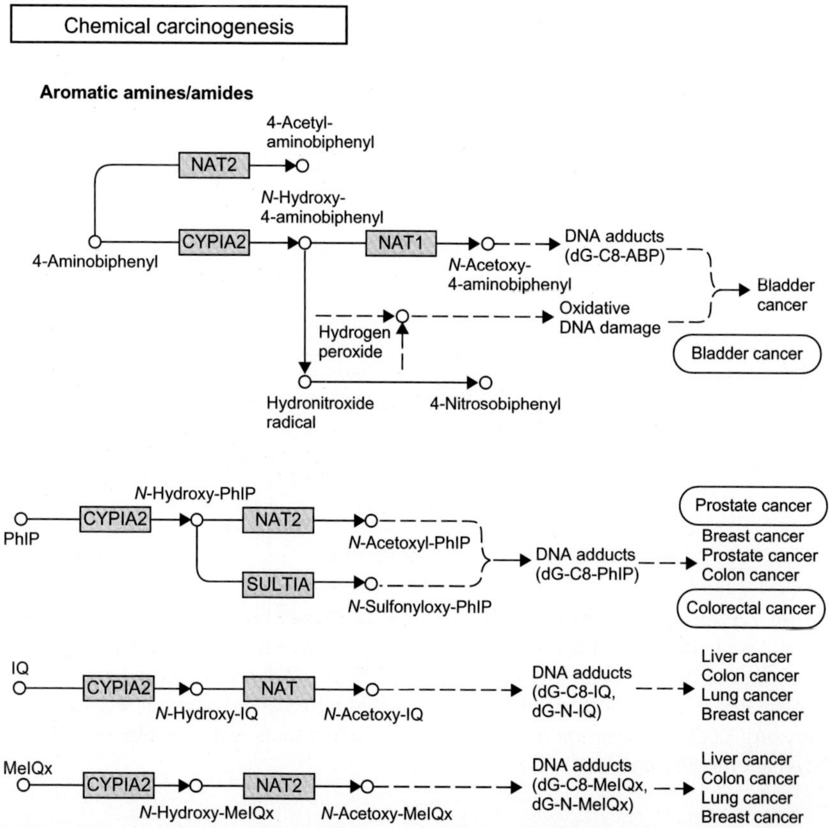

Chemical carcinogenesis

Aromatic amines/amides

Figure 7.4 ***An example of a chemically induced damage in KEGG.*** KEGG contains more than metabolic pathways. Here an example of chemical carcinogenesis is shown. In this example the mechanisms by which aromatic amines and amides are transformed through various enzymatic reactions to form toxic metabolites that lead to DNA damage and adducts and ultimately cancer. That said, the molecular events of carcinogenesis are not included in the pathway. *KEGG*, Kyoto Encyclopedia of Genes and Genomes. *Reproduced from KEGG with permission.*

inclusion of the AOP framework in the Organization for Economic Cooperation and Development (OECD). Originally tasked with overseeing the Marshall Plan in 1948, OECD works with governments, policymakers, citizens, and other stakeholders to establish evidence-based norms and standards for a range of social, environmental, and economic challenges. By supporting the development of AOPs the OECD provides a level of authority on the application of AOPs in environmental risk assessment and regulatory decision-making by agencies tasked with protecting

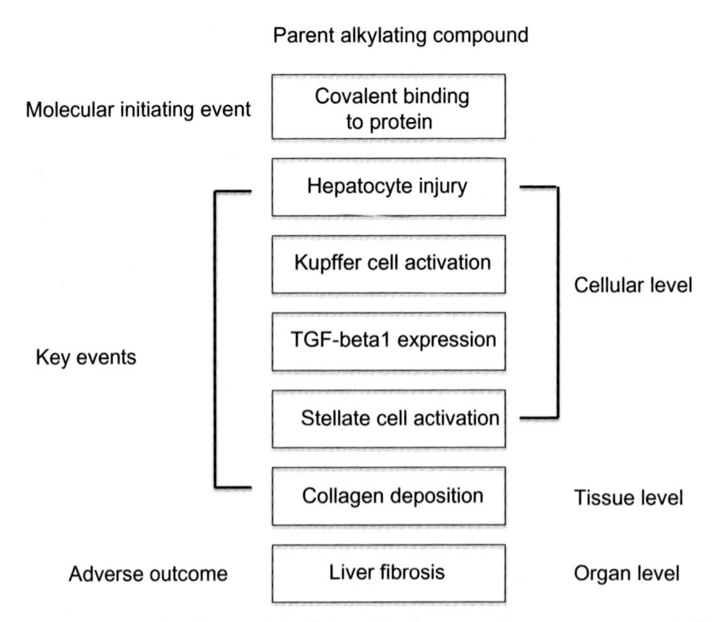

Figure 7.5 ***An example of an AOP.*** This AOP illustrates how exposure to alkylating agents can lead to liver fibrosis (see text for details). This molecular pathway is modified from the AOP provided by the OECD AOP library "Adverse outcome pathway on protein alkylation leading to liver fibrosis" by Bridgette Landesmann at www.oecd-ilibrary.org/. *OECD*, Organization for Economic Cooperation and Development. *AOP*, Adverse outcome pathway.

humans and the environment from chemical exposures. OECD helps establish and maintain standards for this and related projects.

The basic concept of the AOP is shown in Fig. 7.5. One starts with a compound or class of compounds. In the example shown, it is an alkylating agent that can damage DNA. The initial molecular event, or molecular initiating event, is the covalent binding of the alkylating agent to proteins. This is followed by a series of important biological steps or KE, such as cellular injury, inflammatory response, including infiltration of immune cells, and collagen deposition. Together these KE contribute to the adverse outcome of liver fibrosis. Thus, the AOP tracks the effects of the chemical from the molecular to cellular to organ level. Each of the KE has the potential to be a marker of disruption of this biological pathway. AOPs have been developed for hundreds of pathways. This sort of work provides a useful conceptual framework and takes an extraordinary amount of time, development, and validation. It helps regulatory agencies think at a more holistic and systems level, but as far as I know, they are not powered by the sort of power-law functions and differential equations that would allow widespread computational predictions. The reader is encouraged to explore the aopwiki.org website to learn more about how AOPs are developed and how they can help organize our understanding of toxic processes that occur from exposure to environmental chemicals.

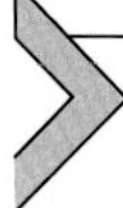

7.5 Systems approaches to understand big toxicological data

Hartung and McBride (2011) urged toxicologists to adopt an approach they referred to as pathways of toxicity (PoTs) which they defined as "a molecular definition of cellular processes shown to mediate adverse outcome of toxicants." These pathways focus on modes of actions instead of the endpoint of toxicity. Analysis of PoTs created the "human toxome," a collection of a finite number of pathways that are involved in toxic response (Kleensang et al., 2014). One of the motivations for their approach was their belief that such a compilation will bolster regulation and reduce reliance on animal testing. These concepts provide a framework that can be used to develop tools that let us leverage the power of networks that help decipher interactions between different layers of biomolecules, as they fall under the central dogma.

Similarly, Hartung et al. describe the need to measure omics profiles as responses to perturbations from chemical exposures (Kleensang et al., 2014). They argue that chemicals can leave a "footprint" in a biological system that can be captured using omics techniques such as metabolomics and transcriptomics that measure the layer closest to the functional response. Furthermore, interactions between different omics layers will uncover regulatory processes and help identify key hubs in biological networks. These hubs can provide information relevant for identifying vulnerabilities as well as possible targets of intervention. This is in line with the biological response component of the exposome definition from Chapter 1, The Exposome: Purpose, Definitions, and Scope. The integration of data obtained from different studies poses further challenges and requires rigorous inclusion criteria. One study analyzed epigenomics, transcriptomics, and proteomics data gathered from 12 cohorts of arsenic exposure. They found common genes and pathways affected across all 12 studies and reported tumor necrosis factor as the top upstream regulator of gene expression changes due to prenatal arsenic exposure (Laine & Fry, 2016) demonstrating the utility of the approach.

7.6 Aggregate exposure pathways

The aggregate exposure pathway (AEP) concept is still emerging and represents an analogous framework to AOPs but focuses on how chemicals are behaving in the environment and as they enter our bodies (Escher et al., 2017). I am not sure if this will be adopted for regulatory decision-making, but conceptually, it is an excellent idea. Some form of the aggregate exposure pathway is essential for exposome research. The AEP framework examines the exposures in combination as they actually occur. Let us imagine a hypothetical city that has only three forms of pollution: auto exhaust, emissions from a coal-fired power plant, and a factory that makes pesticides. Even though each source will emit dozens of pollutants, there should be common compositions from those specific sources. Depending on the composition of the chemicals measured from high-resolution mass spectrometry, it may be possible to determine the source of an individual's exposure. Similarly, using a hierarchical community network model, Li et al. (2019) showed that there were common metabolic responses seen with specific exposure mixtures.

As shown in Fig. 7.6, the aggregate exposure pathway takes into account the composition of the chemical exposures in the environment, including the source. The same high-resolution mass spectrometry described in Chapter 5, Measuring Exposures and their Impacts: Practical and Analytical, can be used to measure these chemicals in soil, water, and air. Sentinel species can also be used to test for the presence of chemicals. These chemicals interact with one another and can result in mixtures that contain not only the parent chemicals, but also products of their associations. Humans are then exposed through some media (or combinations thereof), and the chemical mixtures can then undergo transformations within the body. This can occur at surfaces with microbial components or within the body through enzymatic reactions in the liver, tissues, or cells. Thus, the aggregate exposure pathway framework can help understand what the actual exposures are, including the relative composition of the complex set of chemicals to which one may be exposed. This framework may help integrate information on complex exposures that can inform

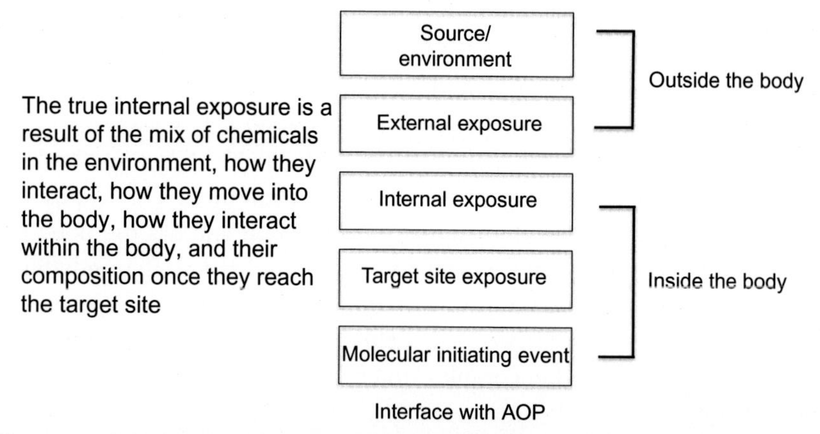

Figure 7.6 ***An example of an AEP.*** The AEP concept has not been as widely adopted as the AOP framework, but it helps illustrate the fact that many exposures occur in nonrandom combinations. Exhaust from automobiles contains a complex mixture of pollutants that are unique to gas or diesel-fueled vehicles. Depending on what is being manufactured, factories also have particular patterns of pollutant emissions. Understanding the network architectures of the exposures can help identify sources and may allow for data dimension reduction strategies to be used. AOP, Adverse outcome pathway; AEP, aggregate exposure pathway.

risk assessment and regulatory decision-making, but it is still in its infancy. However, it is easy to see how the data derived from AEPs and AOPs fit nicely into an exposome-based analysis of human health and can help integrate the information on exposures and biological pathways. I will explore the concept of exposure structure more in the next section on networks.

7.7 Networks: an organizational framework for the exposome

The following section on networks could as easily have gone into Chapter 8, Data Science and the Exposome. Network science is a relatively new field that aims to better understand complicated systems by studying the underlying organizational architecture. Our group examined this in a recent article (Kalia, Jones, & Miller, 2019). Historically, much of the research examining the link between environmental exposures on disease has been reductionistic. Investigators examine single chemicals or small classes of chemicals either in test tubes, laboratory animals, or in human epidemiological studies. More recent studies have embraced a system-based approach to capture the effects of the complex exposures on the complex biology. Systems toxicology has emerged as a subdiscipline within toxicology that borrows tools from the field of systems biology to examine complex biological responses to exposures. The exposome paradigm by its nature uses a system-based approach and has the potential to capture the complex nature of exposures and determine the underlying structure of real-world exposures. Combining systems biology approaches with those of network science can help advance exposome research by helping to synthesize the complex interplay between the environment and biology.

The general public is aware of networks and connections. Some of this was popularized by the game Six Degrees of Bacon, in which players try to connect actors through their coappearance in movies with the goal of tying them to Kevin Bacon. Thus, if an actor had appeared in a movie with Bacon, she would have a Bacon number of 1; if they had acted in a movie with a person who had been in a movie with Bacon, they would have a number of 2 and so on. Although this seems like a silly game, the players are actually analyzing the characteristic complexity of the underlying network structure. Duncan Watts and Steve Strogatz used similar social networks to devise some of the fundamental principles behind network science in their seminal work over 20 years ago (Watts & Strogatz, 1998).

As far back as 1960 it was evident that networks did not exist as random collections of nodes. There was an appearance of small-world networks. Essentially among the galaxy of components, there would be small universes or worlds that would have distinct features and operate as a smaller system (today we refer to these small-world networks as low diameter networks). The World Wide Web has become a major surrogate for network scientists because of the relative ease in studying the network components. Sites such as Google, Amazon, and Facebook are critical hubs that control much of the activity. As we will see, the behavior of networks on the World Wide Web are eerily similar to those in biology.

As technology advances from omics-based approaches to cryo-electron microscopy, improving the resolution at which we can study cellular biology and physiology, we have access to incredibly dense data on the component parts of cellular systems. However, it is becoming increasing evident that we must adopt approaches that rebuild the molecular constituents into models that explain multisystem interactions at molecular and cellular levels, or we are left with a pile of meaningless data. The past two decades have seen a steady growth in the field of systems biology (Voit, 2017). My initial foray into systems biology was based upon the aforementioned collaboration with Voit. The biological networks of dopamine neurochemistry we started to develop over 15 years ago to understand the disruption of dopamine signaling that occurred in Parkinson's disease were comparatively simple compared with the exposome-level work we are now pursuing. The system biology work emphasized molecular networks within a cell rather than the individual components of a cell. Understanding complex systems requires them to be studied at a systems level. This is especially true for environmental contributors to human disease as they exist within a complex environmental setting with temporal, spatial, and combinatorial variability.

The use of high-throughput technologies has led to an exponential explosion of the number of data points derived from a single sample, and approaches to synthesize and interpret these types of data are still developing. Here I review progress in the use of principles of network science to merge the information generated by exposome-related science to better understand the complex interplay between the environment and human biology. The advent of omics techniques generates high-dimensional and unbiased measures of cellular function that demand methods that can organize measures of many cellular molecules into interpretable functional and/or physical structures. Systems biology has leveraged principles from graph theory and network science to achieve this organization (Voit, 2017). Network science provides a framework that can help interpret the behavior of molecules that are correlated, vary in their relationships with one another, and form an organized structure- that is, a network (Barabási & Pósfai, 2016; Newman, 2010). These functional networks can be thought of as a collection of nodes, with a pair of nodes being connected to each other via edges. The edges contain information about the strength, direction, or quality of the relationship between the two nodes. This overall structure of a network is similar across different disciplines, making it possible to borrow computational methods while applying the domain-specific interpretation of results. Many have shown that networks in biology exhibit "small-world" network characteristics, that is, they tend to be scale-free, and not simply exhibit a random distribution. Fig. 7.7 illustrates the difference between random and scale-free networks. This type of organization shows the presence of key hubs (influential nodes)

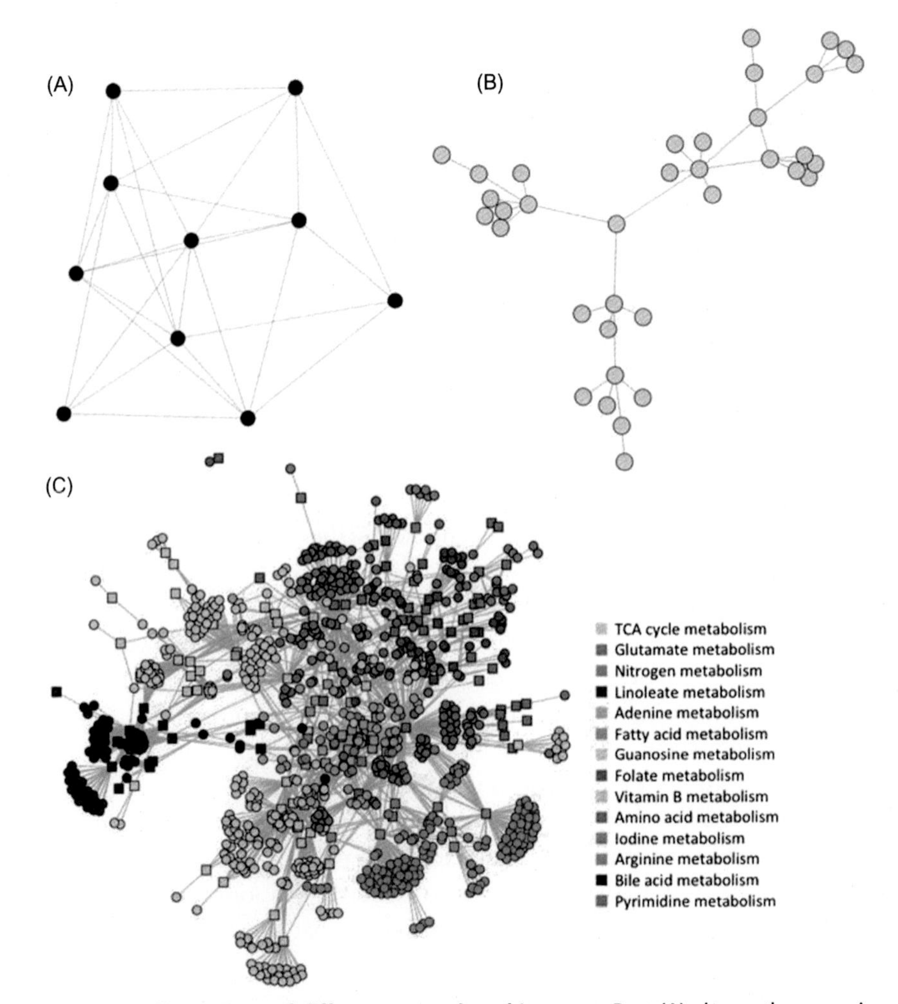

Figure 7.7 ***Illustrations of different network architectures.*** Part (A) shows the organization of a random network where each node has an equal probability of being associated with another. Part (B) shows organization of a protein–protein interaction obtained from yeast recreated from curated data. It is clear that there is a nonrandom distribution. Part (C) shows interactions between metabolites (*square nodes*) and proteins (*circular nodes*) with differential expression in cases of Alzheimer's disease. Differential chemical features were mapped to pathways using mummichog. The interactions between annotated metabolites and proteins were obtained using OmicsNet and graphed using the igraph package in R. The different colors indicate communities as determined by network properties. *Modified from Kalia, V., Jones, D. P., & Miller, G. W. (2019). Networks at the nexus of the systems biology and the exposome.* Current Opinion in Toxicology, *15, 25–31.*

that regulate many downstream processes. These influential nodes can be seen in internet commerce and social media, but they also exist in biological systems. This is not surprising to biologists who believe this affords some level of resiliency through redundancy and emphasizes the importance of key regulators. This organization is apparent in many different biological networks generated using omics data, including protein–protein interactions, transcription factor binding, genetic interaction networks, and metabolic interactions (Fig. 7.7). Thus, network science provides a way to organize large omics datasets to characterize biology and provides a means to understand systemic biological changes that associate with chemical exposures.

7.8 Organized complex exposures?

Human interactions with the surrounding environment are dynamic in space and time. An exposure occurs when chemicals in the environment reach a human–environment interface. The dose of exposure becomes of interest once the chemical crosses the biological barrier and enters the body where it can be circulated. As noted by Hartung et al. (Kleensang et al., 2014), measuring the biological "footprint" of the chemical can serve as a marker of exposure. Importantly, biological responses can be transient or sustained, with the latter contributing to a cumulative and lifelong biological impact, which highlights the need to measure the biological consequences of exposure. Exposure to air pollution can cause an immediate airway reactivity and immune response that resolves over the course of hours but can also cause epigenetic changes that may persist for months to years.

A recent study from the Snyder lab created an interaction network that used personal monitoring data collected over three years (Jiang, Wang, & Li, 2018). Using active air sampling, they collected information on microbiota, several chemical constituents, and plant-derived allergens. They assessed the relationship between different microbiota by creating an interaction network. This type of network was also used to understand the relationship between an individual's exposure to this range of factors. They termed this interaction network a "human–environment cloud" and found it to be diverse, distinct for individuals, and dynamic. Although this study had a small sample size, it serves as an excellent network-based example that integrates various sources and techniques to measure specific components of an individual's exposome (Jiang et al., 2018).

Large environmental datasets produced from untargeted metabolomics/exposomics create complex co-exposure structures that can be organized into interaction networks using correlation information for visualization and characterization. Since exposures co-occur, combinations of exposures can be predicted by the geography, occupation, socioeconomic status, diet, and other lifestyle factors of an individual. This structure can be expanded to include data on disease state to identify clusters of chemicals that are most associated with a disease or biological response. This is an interesting observation that is readily recognized by those studying biological networks, but the idea that the exposures can occur in a semiorganized structure is not typically considered in human exposure studies.

Differentiating exogenous chemicals from endogenous chemicals is a challenge in mass spectrometry-based metabolomics because a large number of the chemical features have not yet been identified, and the fact that it is not clear what investigators mean by exogenous. Furthermore, we do not have the structural framework analogs to KEGG that allows mummichog-type analyses to be applied to exogenous synthetic chemicals. If a chemical is plant-derived and part of the person's diet (and feeds into known metabolic pathways), how does one categorize that? Does it matter from where glucose originates? A more reasonable way to define the exogenous features may be by focusing on those that cannot be made by the body. This becomes more restricted but provides a potentially definable demarcation. For example, our bodies do not naturally make halogenated compounds. Thus, it is easy to conclude that any halogenated species must be exogenously derived (or at least its source). To overcome this limitation, analytical and computational methods have been developed using probabilistic and cheminformatic techniques to measure the exposome. With these approaches, high-resolution mass spectrometry data have been used to link the possible source and identity of a chemical feature with measured exposure, as well as determining the biological relevance of features to measured biological outcomes. For example, mummichog uses Fisher's exact tests to determine which metabolic pathway is most likely enriched based on the differential presence of chemical features (Karnovsky & Li, 2020). This program has been especially helpful in gleaning biological insight that had previously been unattainable when one lacks full identification of features. A new cheminformatic tool

BioTransformer (Djoumbou-Feunang et al., 2019) is a software package that uses machine learning approaches with a knowledge-based approach to predict small molecule metabolism in human tissues and the environment via a prediction tool. These predictions can help in accurate, rapid, and comprehensive in silico compound identification. However, some chemicals can be derived from both sources, generated endogenously and through environmental exposures, posing a challenge in determining the source of such a chemical detected on the mass spectrometer. We have been using a parallel approach to predict xenobiotic metabolites from chemicals subjected in vitro to transformation processes and then analyzing the samples by high-resolution mass spectrometry.

In omics-based experiments, it is recognized that many variables are not independent, that is, they are related to one another due to their involvement in similar synthetic or degradative pathways or having similar spatial characteristics. Thus, exogenous chemicals, whether in the environment or in the body, are likely to exhibit a community network model. Li et al. (2019) examined this possibility in a cohort that had been followed for breast cancer development over multiple generations. The authors identified community clusters around specific classes of compounds, namely DDT and PFAS. By organizing the data into a correlation, network communities of chemicals were readily identified and provided an approach to assessing complex mixtures as is necessary for the exposome (Li et al., 2019). This study provides another example of how network-based approaches can help minimize the complexity of the exposome and make the analytical solutions more tractable.

Many researchers have used the power of high-resolution mass spectrometry in metabolomics to uncover systemic biochemical changes that accompany a disease state (Vermeulen, Schymanski, Barabási, & Miller, 2020). Others have reported the use of inductively coupled plasma mass spectrometry to measure the amount of metals stored in deciduous teeth, which serve as biomarkers of in utero and early life metal metabolic dysregulation, and shed light on mechanisms underlying autism spectrum disorder (Arora, Reichenberg, & Willfors, 2017). While the blood metabolomic profile can reveal metabolic changes associated with a disease process, it is likely confounded by changes in the metabolome associated with the disease process itself. This issue of temporality in assigning cause and effect can be controlled by employing prospective, longitudinal cohort designs in epidemiological studies of the exposome.

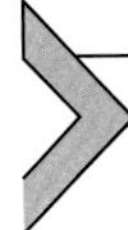

7.9 Integrating toxicological and exposomic data to observe biological response

While techniques that can collect big data have been gaining steadily, so have methods to integrate environmental and biological data. One approach uses quantitative and qualitative measures to describe a tripartite network consisting of interactions among environmental exposures, biological pathways, and human phenotypes to determine the contributions of environmental and biological changes toward a disease outcome (Fig. 7.8). Multiscale, multifactorial response networks also provide an

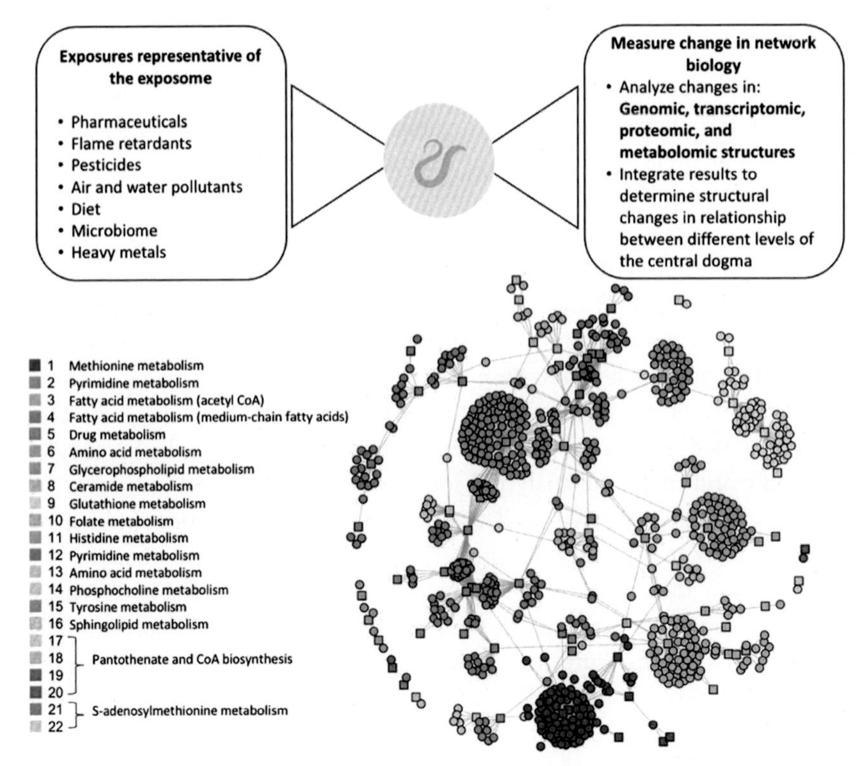

Figure 7.8 ***The exposome in the laboratory.*** Model organisms provide an approach to study the effects of defined exposures of chemicals on changes in network biology. Here, we see changes in the metabolomic network of *Caenorhabditis elegans* due to exposure to a known neurotoxicant, MPP^+. This was conducted in the lab using wild-type worms, exposed to 1 mM MPP^+. High-resolution metabolomics was performed on a Thermo Q-Exactive HFX. *Modified from Kalia, V., Jones, D. P., & Miller, G. W. (2019). Networks at the nexus of the systems biology and the exposome.* Current Opinion in Toxicology, *15, 25–31.*

opportunity to understand the relationship between variables across multiple networks using tools such as partial least squares regression (Li, Sullivan, & Rouphael, 2017). An occupational health study used a combination of occupational exposure data, HRMS, and phenotypic data to show an interaction between exposure to trichloroethylene and markers of renal function. They report the presence of previously unidentified chlorinated metabolites that were strongly correlated with markers of renal dysfunction (Walker, Uppal, & Zhang, 2016).

In Europe, the Human Early Life Exposome study (HELIX) was a prospective study designed to collect biological data and exposure assessments from gestation to early adolescence in more than 30,000 mother–child pairs living in six countries (Maitre, de Bont, & Casas, 2018). This birth cohort is designed to collect omics profiles and exposure measures made using biomarkers and environmental measurements. Although the use of prospective cohort studies in exposome research reduces uncertainties concerning causal association, they are expensive and may still have biases introduced by exposure misclassification and selection bias. Experimental toxicologists have an important role to play in uncovering the relationship between complex biological responses and the exposome. Using animal models, toxicologists can systematically begin to explain the network changes associated with exposure to multiple chemicals. Evolution has conserved various pathways and molecular machineries across species making model organisms great sources of biological information, with human relevance. Using organisms such as *Caenorhabditis elegans* (worms), *Danio rerio* (zebrafish), and *Drosophila melanogaster* (fruit fly), toxicologists and biologists can make important links between exposures and biological outcomes. Since these three animal models have been widely used in biology and toxicology, the curation of many aspects of their genome, transcriptome, proteome, and metabolome is available. Their short lifespan makes it possible to study the lifelong effects of exposure and makes it possible to study aging, lifespan, and health span. We have used *C. elegans* as a model to study omics level alterations in metabolism after chemical exposure (Fig. 7.8). Furthermore, cross-species network analyses can be used to uncover evolutionarily conserved pathways that are perturbed by environmental exposures or an underlying disease process. Apart from model organisms, organoids and organs on chips provide avenues to study cellular function with tightly controlled environmental conditions. These approaches offer opportunities to assess targeted questions that may arise from untargeted, network approaches applied in measuring the effects of the exposome.

Research in environmental health and toxicology is generating ever-increasing amounts of data that point to toxic effects of several environmental exposures. These data have led to policy changes and regulatory implementation that have saved many lives and improved quality of life. However, much of this earlier work was focused on a small number of high-volume chemicals-for example, lead and mercury. There is now a need to take an approach that allows us to assess the impact of the multitude of low-abundance chemicals. Using an exposome framework within the field of toxicology will provide a new level of understanding that considers the complex nature of real-world exposures to hundreds to thousands of chemicals and the corresponding biological responses. This endeavor is dependent on access to high-quality databases that are amenable to computational manipulation. Biosynthetic experiments and cheminformatic algorithms are required to fill databases on environmentally relevant chemicals, which remain sparse and need to include metabolic products of parent toxicants (see the EPA CompTox Chemicals Dashboard—https://comptox.epa.gov/dashboard). Although we still face limitations in our ability to interpret omics data, advances in data curation and generation will improve our ability to identify new links between exposure and response. Integrating systems biology and the exposome using a network science, framework will provide a measure of the environmental basis and contribution to human disease.

I am completely convinced that the future of exposome will require the development and application of network science-based approaches to interpret data from our complex exposures and the associated biological responses. The environmental externalities and biological internalities are best represented in network form and how they change over time require approaches that retain the rich information from within the networks. The data science core (originally the systems biology core) we built under the HERCULES Exposome Research Center at Emory University was created to address the multifaceted data analytics that were needed as exposome research proceeded. I have been on a mission to recruit investigators with expertise in network science, machine learning, and other data science approaches to help analyze exposome-related data, and my relocation to Columbia has given me access to additional expertise. In addition to my efforts to recruit outside expertise to my own research projects, I have been encouraging others to become involved in exposome research at a broader level. The complexity of the exposome is the type of problem data and network scientists want to pursue as they pose robust computational challenges that require the development of new solutions.

7.10 Obstacles and opportunities

The technical challenges of pathway- and network-based science will not be a limitation for exposome research. The methods are being developed for many other fields from physics to finance. The obstacle is the exposome field not having access to the expertise to apply these approaches to our specific problems. As will be emphasized in the next chapter, there is fierce competition for data scientists, and the same can be said for those with expertise in network science (which could be argued is a subset of data science). Thus, the exposome must work to recruit investigators with expertise in network science into the field and at the same time develop training programs that teach students, postdocs, and faculty how to use these approaches to solve exposome-related problems. The recent collaboration with Barabási (Vermeulen, Schymanski, Barabási, & Miller, 2020) illustrates that there is interest in the network science community for exposome-related research, but there simply are not enough exposome researchers with the necessary expertise. Although we are working to integrate network science into our environmental health training programs at Columbia University, this will just impact a handful of trainees. In order to meet the emerging demand, other universities will need to expand their training programs to include advanced data science methods into their environmental health and exposome-related training programs.

7.11 Discussion questions

1. At what point in your education do you think that concepts of network science and systems biology could have been introduced? Do these concepts require advanced mathematical or biological knowledge?
2. How might you incorporate pathway analysis or network science into your work (whether it be a job, research project, or education)?

3. Play the six degrees of separation game with your classmates or work colleagues. Can you identify a common acquaintance on a different continent within six steps?
4. How does one balance detailed knowledge of molecular action of proteins and chemicals with knowledge of biological networks and physiological systems?

References

Arora, M., Reichenberg, A., Willfors, C., et al. (2017). Fetal and postnatal metal dysregulation in autism. *Nature Communications*, *8*, 15493.

Barabási, A.-L., & Pósfai, M. (2016). *Network science*. Cambridge University Press.

Djoumbou-Feunang, Y., Fiamoncini, J., Gil-de-la-Fuente, A., Greiner, R., Manach, C., & Wishart, D. S. (2019). BioTransformer: a comprehensive computational tool for small molecule metabolism prediction and metabolite identification. *Journal of Cheminformatics*, *11*(1), 2.

Escher, B. I., Hackermüller, J., Polte, T., et al. (2017). From the exposome to mechanistic understanding of chemical-induced adverse effects. *Environ Int.*, *99*, 97–106. doi:10.1016/j.envint.2016.11.029.

Hartung, T., & McBride, M. (2011). Food for thought...on mapping the human toxome. *ALTEX*, *28*, 83–93.

Jiang, C., Wang, X., Li, X., et al. (2018). Dynamic human environmental exposome revealed by longitudinal personal monitoring. *Cell*, *175*(1), 277–291.

Kalia, V., Jones, D. P., & Miller, G. W. (2019). Networks at the nexus of the systems biology and the exposome. *Current Opinion in Toxicology*, *15*, 25–31.

Kanehisa, M., Goto, S., Sato, Y., Furumichi, M., & Tanabe, M. (2012). KEGG for integration and interpretation of large-scale molecular datasets. *Nucleic Acids Res*, *40*(2012), D109–D114.

Karnovsky, A., & Li, S. (2020). Pathway analysis for targeted and untargeted metabolomics. *Methods in Molecular Biology*, *2104*, 387–400.

Kleensang, A., Maertens, A., Rosenberg, M., Fitzpatrick, S., Lamb, J., Auerbach, S., ... Hartung, T. (2014). Pathways of toxicity. *ALTEX.*, *31*, 53–61.

Laine, J. E., Fry, R. C., et al. (2016). A systems toxicology-based approach reveals biological pathways dysregulated by prenatal arsenic exposure. *Ann Glob Health*, *82*(1), 189–196. https://doi.org/10.1016/j.aogh.2016.01.015.

Li, S., Sullivan, N. L., Rouphael, N., et al. (2017). Metabolic phenotypes of response to vaccination in humans. *Cell*, *169*(5), 862–877.

Li, S., Cirillo, P., Hu, X., et al. (2019). Understanding mixed environmental exposures using metabolomics via a hierarchical community network model in a cohort of California women in 1960's. *Reprod Toxicol.*, *S0890-6238*(18), 30603–30608. https://doi.org/10.1016/j.reprotox.2019.06.013.

Maitre, L., de Bont, J., Casas, M., et al. (2018). Human Early Life Exposome (HELIX) study: A European population-based exposome cohort. *BMJ Open*, *8*(9), e021311.

Newman, M. E. (2010). *Networks: An introduction*. Oxford, UK: Oxford University Press.

Ogata, H., Goto, S., Sato, K., Fujibuchi, W., Bono, H., & Kanehisa, M. (1999). KEGG: Kyoto Encyclopedia of Genes and Genomes. *Nucleic Acids Research*, *27*(1), 29–34.

Qi, Z., Miller, G. W., & Voit, E. O. (2008a). Computational systems analysis of dopamine metabolism. *PLoS One*, *3*(6), e2444.

Qi, Z., Miller, G. W., & Voit, E. O. (2008b). A mathematical model of presynaptic dopamine homeostasis: implications for schizophrenia. *Pharmacopsychiatry*, *41*(Suppl. 1), S89–98.
Qi, Z., Miller, G. W., & Voit, E. O. (2009). Computational analysis of determinants of dopamine (DA) dysfunction in DA nerve terminals. *Synapse*, *63*(12), 1133–1142.
Qi, Z., Miller, G. W., & Voit, E. O. (2010). Computational modeling of synaptic neurotransmission as a tool for assessing dopamine hypotheses of schizophrenia. *Pharmacopsychiatry*, *43*(Suppl. 1), S50–60.
Qi, Z., Miller, G. W., & Voit, E. O. (2014). Rotenone and paraquat perturb dopamine metabolism: A computational analysis of pesticide toxicity. *Toxicology*, *315*, 92–101.
Vermeulen, R., Schymanski, E. L., Barabási, A. L., & Miller, G. W. (2020). The exposome and health: Where chemistry meets biology. *Science*, *367*(6476), 392–396.
Voit, E.O. (2017). *A first course in systems biology* (2nd ed.). Garland Science.
Voit, E. O., Qi, Z., & Miller, G. W. (2008). Steps of modeling complex biological systems. *Pharmacopsychiatry*, *41*(Suppl. 1), S78–84.
Walker, D. I., Uppal, K., Zhang, L., et al. (2016). High-resolution metabolomics of occupational exposure to trichloroethylene. *International Journal of Epidemiology*, *45*(5), 1517–1527.
Watts, D., & Strogatz, S. (1998). Collective dynamics of small-world networks. *Nature*, *393*, 440–442.

Further reading

Barabási, A.-L. (2002). *Linked: How everything is connected to everything else and what it means for business, science, and everyday life*. Cambridge, MA: Perseus Books Group.
Hartung, T., van Vliet, E., Jaworska, J., Bonilla, L., Skinner, N., & Thomas, R. (2012). *Systems Toxicology*, *1*(29), 119–128. Available from https://doi.org/10.14573/altex.2012.2.119.
Uppal, K., Ma, C., Go, Y.-M., Jones, D. P., & Wren, J. (2018). xMWAS: A data-driven integration and differential network analysis tool. *Bioinformatics.*, *34*, 701–702.
Uppal, K., Walker, D. I., & Jones, D. P. (2017). xMSannotator: An R package for network-based annotation of high-resolution metabolomics data. *Analytical Chemistry*, *89*, 1063–1067.
Villeneuve, D. L., Crump, D., Garcia-Reyero, N., Hecker, M., Hutchinson, T. H., LaLone, C. A., ... Whelan, M. (2014). Adverse outcome pathway (AOP) development I: Strategies and principles. *Toxicological Sciences*, *142*, 312–320.
West, G. (2017). *Scale: The universal laws of growth, innovation, sustainability, and the pace of life in organisms, cities, economies, and companies*. Penguin Press.

CHAPTER EIGHT

Data science and the exposome

The data generated from the exposome defy human intuition.

Eberhard Voit

8.1 Introduction

Given my obsession with definitions, it seems prudent to begin this chapter by defining what I mean by data science. My preferred definition of data science is the one promulgated by my colleague at Columbia University, Jeannette Wing. Wing directs the Columbia University Data Science Institute, was former Vice President of Research at Microsoft, and led the department of computer science at Carnegie Mellon University; thus, she would appear to be well positioned to proffer a definition for such an important and evolving field. Simply stated:

data science: the science of extracting value from data

This simple definition is very powerful and useful for a wide range of investigators. Acknowledging the fact that the words science and data are part of the term being defined and the definition itself, which may make English majors twitch, the definition clearly conveys the overall purpose of the field—namely, extracting value from data. Extracting value underscores the difference between mere data points and value or knowledge. The word value is subjective and dependent on the particular domain or user, which is appropriate, as it allows the context to be incorporated as needed. The term "big data" does not suggest anything of value; it just represents a massive collection of bits of information. Unsupervised machine learning and big data may make for fun computational exercises, but it more often than not fails to deliver value. Data science, though, is much more than machine learning; it includes rigorous study design, articulation of the objective or question, careful curation of data, assurance of data quality, an armament of advanced analytical tools, and steps to ensure reproducibility and reliability. It is this suite of approaches that

The Exposome.
DOI: https://doi.org/10.1016/B978-0-12-814079-6.00008-0

help deliver the value. Data science has emerged as a unique science or defined discipline. There is a community of data science scholars working on the development and validation of methods, and an even larger community taking advantage of both longstanding and emerging tools and techniques. Exposome research is going to generate an extraordinary amount of data, but if we do not build systems, train scientists, and apply the best methodology for the field, we may fail in gaining knowledge. This is why data science is so critical for the exposome.

Another quote that is apropos for this chapter is one I have heard John Quackenbush, Chair of Biostatistics at Harvard Chan School of Public Health, cite on a couple of occasions:

> *Every revolution in science—from the Copernican heliocentric model to the rise of statistical and quantum mechanics, from Darwin's theory of evolution and natural selection to the theory of the gene—has been driven by one and only one thing: access to data.*

John uses this quote to illustrate the essential and inarguable symbiotic relationship between data and science. Data are the scientist's intellectual currency. John searched in vain for a quote to capture the spirit in the above mentioned words, and when he failed to identify one, he did what any entrepreneurial thinker would do—he made it up. Thus, it is important to give this biostatistician, who was trained in theoretical physics and worked on the Human Genome Project, proper credit for this insightful, field-spanning, and Kuhn-invoking quote that illustrates the intertwined nature of data and science. The aforementioned quotes illustrate that the generation, curation, and analysis of data have been and always will be part of science. Every theorem and hypothesis relies on data for formulation, confirmation, or refutation. For centuries, classical, mathematical, and statistical methods were able to handle the volume and variety of data; however, as science has advanced, the demand for more sophisticated approaches helped create a new field dedicated to the data ecosystem. It just so happens that the exposome is coming of age in the same era and benefits from growing up as part of the same generation as data science.

8.2 Of grave importance

As I noted earlier, as we developed the HERCULES Exposome Research Center we systematically assembled expertise in data science, including experts in database architecture, biostatistics, machine learning, bioinformatics, systems biology, computational toxicology, and cloud computing. As the exposome field grows, I am seeing additional needs in areas such as network science and data visualization, and these are areas I am now working to incorporate into our exposome efforts at Columbia University. The range of data science expertise needed for the exposome is substantial, and my sense is that this may be a bigger hurdle than the technological and mass spectrometry–based challenges we face. I doubt it is a limitation of the data science technology itself, but more the difficulty in attracting the best data scientists to specifically work on exposome-related problems and projects.

I am not an expert in data science and do not proport to be a data scientist; however, I have long recognized the need to have access to advanced computational and bioinformatics tools and have formed numerous alliances with experts willing to collaborate on my scientific forays. I state this to ask the statistically sophisticated reader for forgiveness if I have oversimplified or simply misstated much of what follows. I like to consider myself an observational statistical or mathematician. By that I do not mean I use observational data, but rather I like observing other people wield their mathematical prowess to solve complex problems. As noted in the previous chapter, my collaborative efforts with the mathematician and systems biologist Eberhard Voit and my following of the network wizardry of Alex Vespignani and László Barabási have given me a deep appreciation of mathematical and computational thinking as it relates to the impact of complex exposures on health. This has even bled over to my more leisurely pursuits. During my visiting professorship at the University of Paris Descartes, my apartment faced the Montparnasse cemetery. As I strolled through the cemetery (because this is the type of thing that one does when one is a visiting professor), I came across the tomb of Henri Poincaré (Fig. 8.1), the French mathematician who proposed a theory on the shape of the universe. (I am not purely a science geek, I also made a sojourn to Père Lachaise to see Jim Morrison's and Chopin's graves.) The resulting Poincaré Conjecture stymied the field of mathematics for a century. Just a few years ago, an obscure and reclusive mathematician from St. Petersburg, Grigori Perelman, provided the long-sought-after solution. There are several unique features of the story of the solution: the sheer ingenuity of the solution, the use of the preprint server arXiv for posting his solution, and his refusal to accept the Field's Medal for his achievement. The solving of the Poincaré Conjecture is considered

Figure 8.1 Tomb of Henri Poincaré. The Poincaré Conjecture perplexed mathematicians for nearly a century. The solution by Grigori Perelman is considered one of the greatest mathematical solutions of all time. Montparnasse Cemetery, Paris, France, 2018.

one of the great mathematical achievements of all time, and in fact, some consider it one of the great intellectual feats of all time (Shea, 2007). The exposome is not in need of a singular Perelman-type genius solution, but rather a series of sustained data science–driven advances over many years that will require a legion of mathematically-inclined investigators. We must devise a strategy to recruit, cultivate, develop, and incentivize the analytical team. I emphasized the need for collaboration with geneticists in the opening chapters, and I believe that many of the talented investigators who have helped solve genomic problems may be key to exposome solutions, especially the mathematically-inclined statistical geneticists. We likely need the analogous statistical experts for the exposome (but the term "statistical exposomicists" should never enter the scientific vernacular).

8.3 Computational and causal thinking

The classical scientific method involves observation, hypothesis generation, experimentation, additional observation and data collection, and drawing conclusions from the observed data. This linear, yet iterative, process is taught in introductory science classes. When one is able to manipulate independent variables and control other conditions, it can be fairly straightforward to test a hypothesis. This works especially well with test tubes and purified chemicals or pathogens, but not as well in human populations. Even setting aside ethics for a moment, it is not possible to readily manipulate independent variables as they relate to the exposome in human populations. For many years, I avoided human studies because of this limitation. However, as I learned more about causal inference and randomized control trials, I started to understand that there were approaches to studying causation even when one is unable to control or manipulate all of the variables. Causal inference can be defined by the process of determining whether or not certain conditions or events directly lead to a change in a dependent variable; essentially it can be considered the study of cause and effect when one is unable to manipulate the independent variables as can be done in the laboratory.

In Judea Pearl's *The Book of Why* (Pearl & Mackenzie, 2018), the authors chronicle the field of causal inference (as well as Pearl's own journey in the field). I am introducing the concept of causal inference early in this chapter to make a critical point. Merely having large amounts of data will be meaningless for the exposome. We must have a causal framework mentality even if the initial experimentation is unable to address it directly

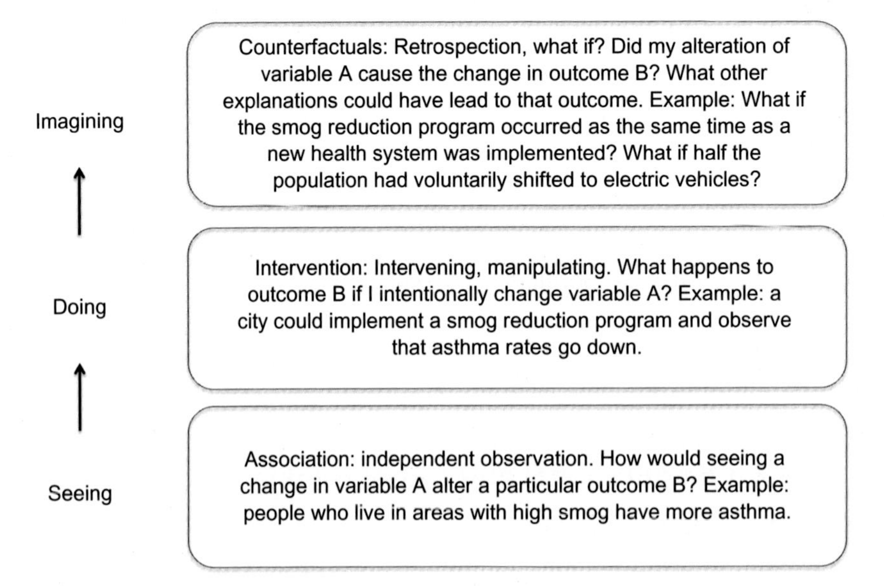

Figure 8.2 Pearl's Ladder of Causation. Pearl describes the steps in causation as a ladder. As one goes up each rung of the ladder a stronger causal connection is made. This figure illustrates the steps of association, intervention, and counterfactuals described by Pearl (Pearl & Mackenzie, 2018).

(Fig. 8.2). It is not that every exposome-related study has to have a causal framework, but rather the long-term goal should be about causation that is grounded in biology. In another nod to genetics, it was the work of the geneticist Sewall Wright who helped lay the groundwork of causal inference through his study of a model slightly more advanced than Mendel's peas—guinea pigs. The unusual coat color inheritance pattern of the animals allowed him to develop causal maps to calculate their mottled coat color, as well as insight into birthweight and gestational time (Pearl & Mackenzie, 2018). My desire to have exposome-level data in nearly every biomedical research study is not about just having more data. The variety of studies will help create and interrogate the causal network. There are excellent examples of using causal methods in environmental epidemiology that provide a model for these types of approaches (Schwartz et al., 2018). When we see examples of people's environments changing due to job relocation or natural disasters, we start to build databases that resemble randomized clinical trials. When you get data on a million people who have been treated for diabetes, you can gain great insight on the best treatment strategy, but if you also have exposome-level information (external exposures, biological measure, and spatial information) on that same population, you could start constructing the causal diagrams that allow one to make conclusions about how the environment impacts disease progression and responses to treatment. It is possible to integrate data on drug response and disease progression with genetic data, even if it was not collecting during the study (you can go back and genotype). But getting exposure-level data requires integrating this data collection at earlier states (or at least biological sample collection that can be interrogated later).

Historically, laboratory scientists are taught the art and science of reductionism. The goal is to dissect out the finest details to uncover molecular mechanisms. Many a Nobel prize has been awarded for this type of discovery. Yet, the mysteries of the exposome are much more likely to come from a systems-level approach. This is where the exposome scientist can benefit from computational thinking. Specifically, it will be beneficial to think like a computer scientist, as we unravel the complexity of the exposome. In addition to Wing's powerful, yet pithy, previously given definition of data science, I also find her influential essay "Computational Thinking" to be especially insightful (Wing, 2006). This essay has been cited thousands of times and is credited with embedding computational thinking well beyond the field of computer science. Although this essay was focused on using computation-related thought processes in a variety of life processes, it is a striking parallel to what the field of systems biology attempts to do, which is not surprising since much of systems biology is computational in nature. It is especially challenging for the scientist trained in reductionism to relinquish data processing and problem-solving to a computer, but if one understands the architecture of the underlying algorithms it becomes easier to see how the processing speed of computers serves as a complement to classical intellectual analysis. Computers are designed to do what we program them to do. If we want them to make logical decisions, we need causal models. Indeed, Pearl posits that when causal inference is fully integrated into machine-learning algorithms, the machines will be able to pass the Turing test—the inability to distinguish an interaction with a computer from that of a human (Pearl & Mackenzie, 2018). Ironically, when one thinks computationally, one must break down the underlying algorithmic processes into their component parts in a reductionistic manner, but the understanding of the underlying steps then allows one to scale up to study complex systems.

8.4 Making big data manageable data

As we were building our exposome research program and recruiting more data experts, Gari Clifford, now Chair of Biomedical Informatics at Emory, suggested that I read a book. I was expecting a book on Python programming or network science given his expertise, but instead it was a book that proposed a contrarian view of data science called *Data Smog* (Schenk, 2007). Written in 2007, which is old by data science standards, the author presents a compelling case that big data can do the opposite of what it is intended. Rather than improving our understanding, the sheer volume of data overwhelms our mental frameworks and pollutes our ability to make sense of incoming data streams. It may remind the reader of the longstanding tension between signal and noise but makes the case that when the volume of data increases as it has, it can become impossible to see the signal above the cacophony. It does provide some advice on reducing the smog, but the take-home message is that large volumes of unstructured information can have effects on our thinking and decision-making as detrimental as air pollution and smog can have on the health of our lungs. Thus, we must be strategic in the data we plan to generate. We should not blindly generate omic-scale data without dedicating effort to how we will organize, curate, and analyze the resulting output.

One area of research that is attempting to fight the smog is data dimensionality reduction. The idea of this approach is that many of our data streams are related and codependent. Thus, knowing two of the key components may tell you all you need to know about the 12 items being measured. This appears to be scalable to exposome-type data. If you have data on environmental exposures that include hundreds to thousands of chemicals species, it may be possible to use data dimension reduction techniques to focus the data on the primary drivers of the effect. Essentially, these approaches identify the most prominent drivers of the outcomes one is studying. We recently published a paper that explored three different unsupervised dimension reduction techniques (Kalia et al., 2020) that could be of benefit to exposome research. These included principal component analysis, factor analysis, and nonnegative matrix factorization. The point was to illustrate that there are ways of taking the thousands of data points that are generated from untargeted approaches and distill them down to a set of features that may be more amenable to linking with other data streams. There are many other approaches for this, but the idea is that there are strategies to make exposome-level data more manageable. These reduction techniques are utilized on the back end, but we also need to consider how we can reduce the unnecessary data on the front end. This may require a forensic examination of pilot exposome studies to see how to optimize data collection and analyses as the studies increase in size.

8.5 Representations of reality

Maps and models are abstract representations of reality. A map of a particular country is a representation of a country's physical features. The map provides depictions of geography, railways, highways, population density, and many other features. The map is not the country, but it provides incredibly useful information to the user, and that information can be used to navigate the country. As illustrated in the previous chapter, models of biological pathways are merely representations of complicated systems. A model of Kreb's cycle tells the user how the various constituents are modified as they move along the biochemical highway. It provides information about enzymes, cofactors, substrates, and products. The model can help the scientist navigate the biological landscape. Thus, from a thematic standpoint, maps and models are similar in their ability to help make the complex understandable, and both will be covered in this chapter. While many of the topics discussed may seem technologically daunting and potentially far-fetched, the reader is to be reminded of the extraordinary pace of technological advancement.

The ability to geocode patient and exposure data provides an intriguing opportunity to generate interactive maps of environmental exposures. Mapping data-rich exposome information will be beneficial to public health professionals and researchers. This information could also be very useful from a regulatory standpoint. By providing specific information on where particular exposures occur and their proximity to the suspected source(s), such approaches greatly expand our understanding of the dynamic nature of exposures. These types of approaches also begin to capture information about the impact of weather patterns on actual exposure. A person may exercise at a park one mile from a factory, but the exposures could fluctuate dramatically based upon their level of exertion and whether or not it has rained recently. However, all of this information could be detected readily with a really smart smartphone (see Box 8.1).

BOX 8.1 The smartphone—a portable data machine

The year 1984 provides an example of the rapid advancement that has occurred over the past few decades. For the more seasoned scientist, *1984* conjures up images of a dystopian society from George Orwell's novel, a fate we have managed to avoid. For the mid-level scientist, it may jar memories of an Orwellian television commercial that appeared that same year (if you were born after 1984, you should probably read the book *and* watch the commercial). The television advertisement showed throngs of people with shaved heads and drab clothing silently watching a video screen of propaganda (a representation of a scene from Orwell's book). In stark contrast to the drab scenery, a young and brightly dressed female runs past guards and throws a sledgehammer through the screen, disrupting the propaganda film in hopes of waking the masses. The message was that Apple Computer was not going to let 1984 be like Orwell's *1984*. It was the audacious introduction of the Mac. In retrospect the audacity may have been understated. Coincidentally, 1984 was the same year when the first cellular telephone became commercially available. Weighing in at a whopping 2 lb and at a cost of $4000 (nearly $9000 when adjusted for inflation), the Motorola DynaTAC8000X was a far cry from what we have now. Referred to as the brick (its approximate size), the DynaTAC sported 30 min of talk time with 8 h of standby time. After steady and incremental improvement in cellular phone technology, the iPhone was introduced in 2007 and transformed the industry (if you owned an iPhone in 2007 or 2008, think back to the comparative lack of power and memory). No longer was the cellular phone a mere communication tool. It was a portable data management system. The technological power contained within our smartphones is staggering when one thinks back to what was available 30 years ago (at that same time I was learning to use computers on a Commodore Vic20 with 64 kB of memory and a Tandy TRS-80 with a mere 4 kB of memory). Checking one's Facebook or Instagram status while walking to class may not change the world, but the potential to track one's movements along with a host of personal environmental variables (temperature, humidity, particulate matter, various gases, radiation, and luminosity) and biological endpoints (body temperature, heart rate, and sweat composition) could revolutionize spatial environmental exposure assessment. All of these technologies are essentially possible today, and every person could eventually become a sentinel for dozens of key environmental indicators with their data uploaded and mapped in real time. Things that seem impossible now may well be common within our lifetimes. We should prepare ourselves for revolutionary technological advances as discussed in Chapter 6, Innovation and the Exposome, and not be restrained by our current technological framework.

The utility of mapping chemical use and exposures is not new. The field of remote sensing is grounded in geospatial analysis. Numerous organizations have been mapping exposures for decades. For example, the U. S. Geological Survey has generated detailed maps of pesticide use over time. It is important to note that in this example, these are use maps, not exposure maps. Data for hundreds of pesticides are available over many years (see Fig. 8.3). The U.S. Geological Survey data come from the Environmental Protection Agency registry database. The data are provided at a county level, and thus, have relatively low resolution. At a country level, the data are visually impressive, but one quickly realizes that these depictions of seemingly catastrophic exposures do not necessarily translate into human exposure or alterations in human health. The maps tell us about the use of this chemical and reveal where most of our agricultural activity occurs. While these maps are visually appealing, their resolution may be insufficient to draw meaningful conclusions regarding the exposome for individuals. Higher resolution maps will be necessary. Beate Ritz at the University of California, Los Angeles, has generated and validated more extensive maps for the California Central Valley. Her data include distance between the actual pesticide application, based on data from the California Environmental Protection Agency (CEPA), and workplace and residence with detailed geocoding. Importantly, the group also verified the utility of their models by measuring the levels of selected pesticides in the blood of hundreds of workers in those areas (Costello, Cockburn, Bronstein, Zhang, & Ritz, 2009). The results showed excellent concordance with the predictive models. However, these data are based upon the reported purchase (and the presumable use) of currently available pesticides and not the actual levels in the water, soil, and air. Banned compounds would not be reported in the CEPA data, but such compounds were identified in the biological samples, emphasizing the importance of including banned, yet persistent, and unknown chemicals in our analyses. Not only are they indicative of historic exposures, they may be driving adverse health effects. Analysis of current pesticide use may just be a proxy for past exposures to more toxic chemicals. What really matters, though, is how much of these pesticides are getting into our bodies. This is where it becomes necessary to measure specific chemicals in human samples. There is likely to be a significant degree of concordance, as shown by the Ritz group, but we cannot make that assumption. We must conduct the experiments on a large scale and measure a wide array of chemicals in the human body. The exposome should certainly build upon these existing

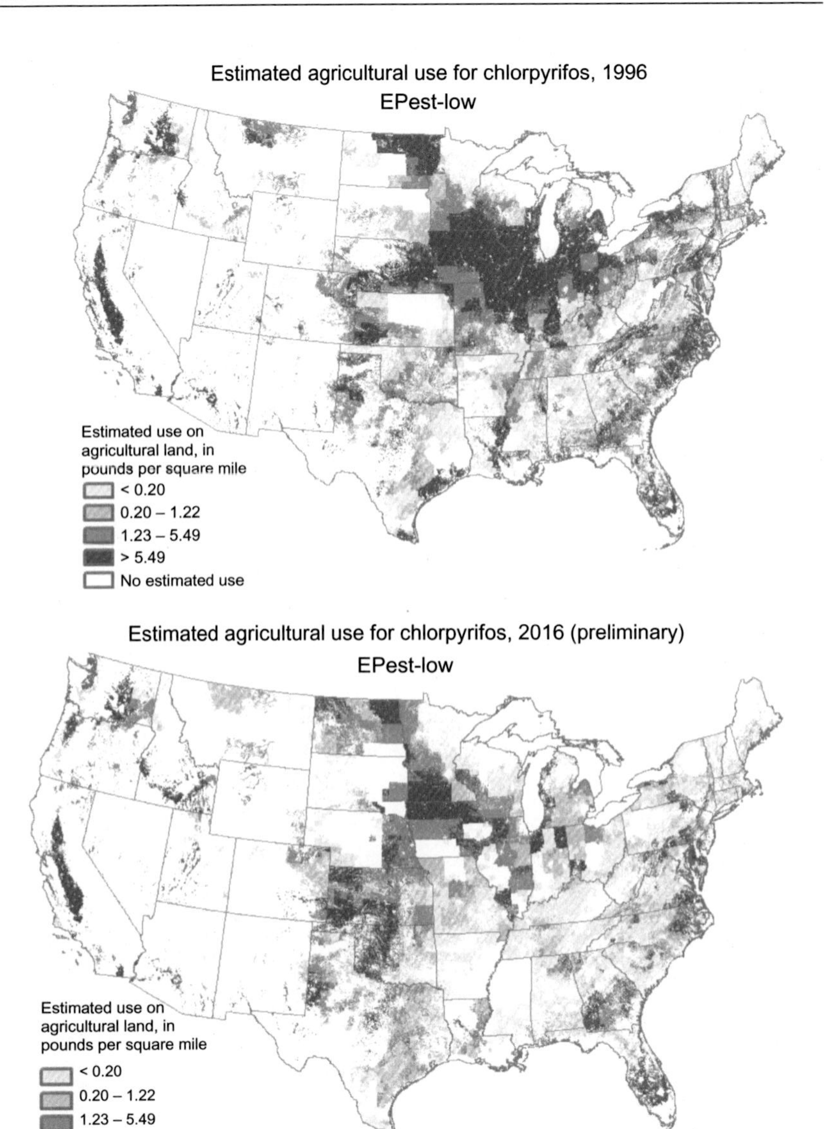

Figure 8.3 Sample map from the U.S. Geological Survey Pesticide National Synthesis Project. As an example, the distribution of insecticide is shown for the years 1996 and 2016 (https://water.usgs.gov/nawqa/pnsp/usage/maps/show_map.php?year = 2016&map = CHLORPYRIFOS&hilo = L).

datasets when possible, overlapping them with census-type data for population, education, socioeconomic factors, and disease incidence, but it is critical to weave this information with quantitative assessment of exposure from humans.

Paul Juarez at the University of Tennessee has been using maps to study environmental health disparities in the context of the exposome. With a policy-based focus, his group is focusing on the data that are currently available to better understand disparities in health. By amassing 30 years of data on environmental factors, such as air and water quality, emissions, toxic waste sites, and more, his group can integrate complex exposome-scale data to examine disease disparities (Juarez, Hood, & Rogers, 2017). The database also includes information on the built environment, housing, social factors, and regulatory factors (zoning, state bans, etc.). His work has used a map-heavy approach focused on the population level to better inform public health.

There are many other potential uses of maps. The traffic flow data on your smartphone comes from information from thousands of smartphones moving along the highway. This technology can provide data on mobility and health. A truck driver's phone is going to log 10 h of moving at 50 mph, which is too fast for exercise and suggests that the user is actually sedentary (driving). A jogger may log 30 min/day at 6 mph. An office worker could spend 10 h/day at 0 mph (at desk). A construction worker may spend 10 h ranging from 0.1 to 1 mph as he or she moves around the work site. Each of these activities (driving, exercising, and working) affects health in different ways, and the generation and collection of these data are relatively simple. It should be possible to automatically determine the type of activity without any input from the user (a bin of 10 mph activity, while possible on foot, would be excluded as physical activity if it was flanked by bins of 50 mph activity, as it is likely a determinant of traffic). It would just take the merging of some of the technologies currently used on GPS and mapping programs on smartphones or on activity trackers such as Fitbit. These streams of data do introduce some interesting ethical challenges further discussed below. We may not want this type of information being sent to our car insurance company as it could also be used to determine reckless driving and speeding (vector patterns should be able to discern driving patterns). Similarly, a health insurance company or employer could be interested in visits to fast food restaurants or bars. This type of information could be helpful in studying health outcomes, but privacy concerns must be addressed. Regardless of how these spatial data are used, it is clear that they will generate massive amounts of data and some of the issues surrounding that will be discussed next.

8.6 Big data, really big

Years ago, most people could ignore the concept of big data, leaving it to the statisticians and quants. However, big datasets are infiltrating our everyday life; from health-care records, to government retirement plans, to social media, and to how stores collect our information and anticipate our desires (marketing on steroids). Big data and its analysis and interpretation have major impacts on our lives and it is only going to become more important. The growing amount of data is becoming a critical issue in biomedical sciences. Each of the omic technologies is generating massive amounts of data, and traditional means of analyses are insufficient to handle it.

In addition to the omic activities, the aforementioned mapping approaches also generate incredibly large datasets. The ever-expanding datasets must be integrated into organizational frameworks. Computational and systems biology approaches will be essential for such an endeavor. Many organizations recognize these concerns and are working on solutions. The European Commission has invested in efforts to "Towards a Common European Data Space" (https://ec.europa.eu/digital-single-market/en/news/communication-towards-common-european-data-space). The U.S. NIH also established Centers of Excellence under the Big Data to Knowledge program (although support for their training programs has dwindled). Some of the approaches mentioned in Chapter 5, Measuring Exposures and their Impacts: Practical and Analytical, are already working with terabyte to petabyte datasets. As we merge the data from the various omics, it will certainly push us beyond the exabyte into the range of the head-spinning zettabyte, yottabyte, and my absolute favorite dinosaur-inspired brontobyte (Fig. 8.4).

While the challenges facing the exposome are not yet part of these big data initiatives, we can certainly take advantages of the tools and approaches that are spurred by these projects (and I predict in the very near future that exposomics will indeed be part of these platforms). New exposome data must be amenable with the systems and approaches that are being developed for the genomic and proteomic initiatives, as well as for the efforts in clinical phenotyping and electronic medical records (EMR). Indeed, a major goal for exposome enthusiasts should be to develop partnerships with these big data centers that are emerging for other fields.

Number (bytes)	Abbreviation	Name
1000	kB	Kilobyte
1000^2 or 1,000,000	MB	Megabyte
1000^3 or 1,000,000,000	GB	Gigabyte
1000^4 or 1,000,000,000,000	TB	Terabyte
1000^5 or 1,000,000,000,000,000	PB	Petabyte
1000^6 or 1,000,000,000,000,000,000	EB	Exabyte
1000^7 or 1,000,000,000,000,000,000,000	ZB	Zettabyte
1000^8 or 1,000,000,000,000,000,000,000,000	YB	Yottabyte
1000^9 or 1,000,000,000,000,000,000,000,000,000	BB	Brontobyte
1000^{10} or 1,000,000,000,000,000,000,000,000,000,000	GB	Gegobyte

Figure 8.4 The impending brontobyte. Current experiments that evaluate environmental exposures are already generating terabytes of data. When these datasets are combined with those from genomic, proteomic, and epigenomic studies, we will undoubtedly be looking at exabytes and zettabytes of data, and it will only grow.

One of my favorite scientific essays is Chaos in the Brickyard by Forscher (1963). In the parable, builders are the metaphorical scientists. Initially, the builders crafted their own bricks to their exact specifications. Later, they employed junior scientists to serve as brick makers to expedite their work. With proper supervision, high-quality bricks continued to be made. At some point the craft of brick making became as valued as the craft of building itself, and thus a new industry developed for the task of making bricks. Bricks were churned out at an ever-increasing pace without effort being put into the actual construction of buildings. There were too many bricks and too few buildings. Eventually, the ground was covered in piles of bricks making it impossible to construct new buildings. The essay was written over 50 years ago. Watson and Crick had just been awarded the Nobel prize. The writer could not have imagined the current state of the brickyard. Given that so much information relevant to the exposome is already present in the pile of bricks, will it be possible to find a clear spot to establish a foundation? It is not clear how a space can be cleared to start building the exposome, but it seems obvious that the blueprints, foundation, and framing must involve computational approaches. There are just too many bricks.

8.7 I am so smart, S-M-R-T

In "*Thinking. Fast and Slow*" Daniel Kahneman stated, "Humans are incorrigibly inconsistent in making summary judgments of complex information." A Nobel Laureate in Economics may seem like an odd source for insight on the exposome, but Kahneman's statement highlights the need for computational systems in the evaluation of the exposome. The exposome is complexity at its finest. Humans are notoriously poor at evaluating large amounts of data and making accurate and consistent decisions. Our brains are wired to simplify complex problems into apparently similar, but more manageable problems, often referred to as heuristics. What we refer to as intuition is actually complex associative processes occurring subconsciously in our brain. It is an impressive level of computation, but it is also fraught with bias and irrational decision-making. We are lulled into a false sense of security and think that we are making wise decisions and sound predictions, but in fact, we fail miserably. Just look at findings from studies of prognosticators across disciplines. Even professionals such as stock traders, political pundits, and business experts rarely perform above statistical chance in the long term (and arguably get paid better than their performance warrants). Many of these complex systems tend to defy predictions anyway, but somehow, we continually believe we can do it with aplomb. Much to our chagrin, cold and robotic algorithms consistently outperform human intuition in medicine and other important decision-making processes. When dealing with the exposome, human intuition will not be sufficient. We must come to grips with our shortcomings and be ready to place our cautious trust in machines, lest we end up with Homer Simpson—like declarations of our intelligence-S-M-R-T.

8.8 Mathematical modeling

It is likely that petabytes of exposome-related data will accumulate before we have had the chance to build a system in which to store and organize it. What is even more intriguing is that so much exposome-related data are being collected even before the overarching and organizing concept of the exposome has been fully embraced. It is not as if there is an alternative concept to organize the exposure-related data. I argue that scientists must take a step back and work on constructing the foundation for such a project. It may be necessary to construct small buildings of focused relationships while we figure out what the neighborhood block will look like.

Given the past, current, and likely future states of computational and systems biology (recall cell phones were not in use until 30 years ago), I would argue that the initial plan should be unrestrained: a truly idealistic and optimally functioning model. For example, a model can quickly assimilate data collected in real time from the individual, such as activity, behavior, food intake, and personal environmental sampling, along with data from somewhat invasive analysis of biological samples for metabolomic, (epi)genomic, transcriptomic, and untargeted chemical analyses. These data could then be entered into the established exposome model to provide a personalized risk profile for thousands of diseases, along with a plan of how to improve the exposome profile.

Such a deliverable may take many years to arrive, but the goal is to provide the motivation and inspiration for the computational technology, approaches, etc. that will need to be developed. This list of desirable tools could serve as a driver for innovation within the field, or could ready the field for rapid adoption into the exposome model once it becomes available. The mega-model may take decades to complete, but within its framework, there will be great opportunities to develop some of the small models that can become part of the larger system and provide the organizational system to contextualize new findings.

It is highly likely that the modeling of the exposome is going to be a multimodal design, but it should be designed in a manner that allows integrations of the various parts. For example, one model may focus on the physicochemical interactions of chemicals in our immediate environment, while another may focus on the transport of those chemicals into the body and the subsequent kinetics of the mixture throughout the body. Others may focus on biological effects at an organ systems level, with each organ or its major functions being modeled. Many of these functions result in alterations in metabolites that reach the bloodstream, for example, reaction products from liver metabolism, chemicals reclaimed or excreted from the kidney, or altered glucose from pancreatic activity. Thus, an exposome model that incorporates the metabolic features found in the blood could capture many of the biological functions occurring within our bodies.

Computational approaches have been used to develop pharmacophores and toxicophores (the models of binding site interactions for drugs and toxicants), and this is how some of the Tox21 data are being modeled and analyzed (Krewski, Andersen, & Tyshenko, 2020). The idea is to use mechanistic toxicology and structural biology to figure out the key properties and then to build computational models that can predict the likelihood of unknown chemicals affecting the target. These computational models can then be used to study the hundreds of chemicals present in the human body to determine the potential additive and synergistic effects. For example, we know that organophosphate insecticides inhibit acetylcholinesterase. We further know that dozens of these products are on the market. If a person has 24 different organophosphates in their bloodstream, should we not evaluate them as a group? Such approaches have been used to some degree at the regulatory level, but it will require more sophistication to make connections between these molecular interactions and disease states. Addressing these problems will require the use of computational models.

8.9 Models

Developing a model is a complicated process. The modeling process has several major steps. An example of how such a model could be developed is shown in Fig. 8.5. The first step is to identify the goals of the project, including the scope. While much of the discussion of the exposome has been boundless, limits will need to be identified when constructing models. What is it we want to model? Are we focusing on exposures first and addressing the various omic projects later? Or do we start with the wealth of genomic, transcriptomic, and proteomic data that are already available? These are not simple questions, but framing the question is a critical step and one that should be carefully decided. Once the goals are identified, it will be necessary to identify the currently available data and what we already know about how the environment impacts health. The next step is selecting the type of model. This decision will be based on a series of issues surrounding the type of data that are presently available and the type of data that is expected to become available. Since time is an important variable, whether it be diurnal variations, day-to-day alterations, or yearly to decade changes, the model will need to be dynamic. Some components may be mechanistic, for example, in a biochemical pathway, but some may correlative, as in how an alteration in a biological pathway is related to a health outcome without knowing why. A regression-based model can make reliable predictions but offers no mechanistic insight. While a multiscale model that could model molecular

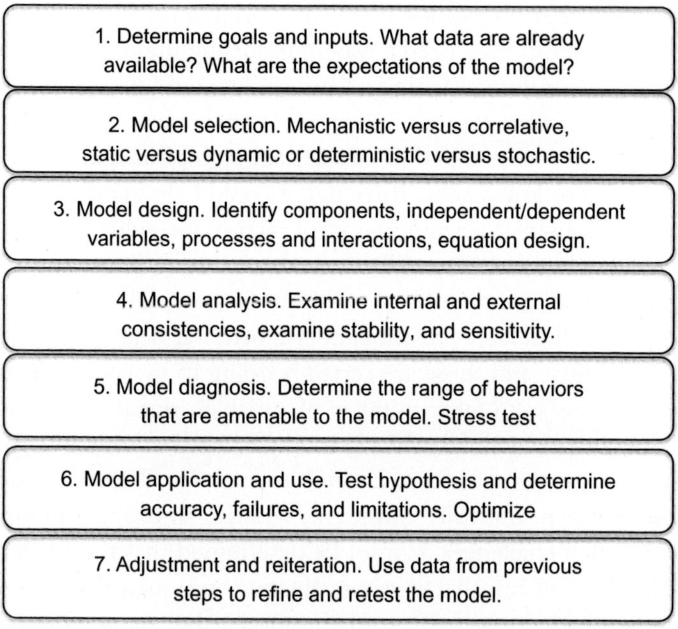

1. Determine goals and inputs. What data are already available? What are the expectations of the model?

2. Model selection. Mechanistic versus correlative, static versus dynamic or deterministic versus stochastic.

3. Model design. Identify components, independent/dependent variables, processes and interactions, equation design.

4. Model analysis. Examine internal and external consistencies, examine stability, and sensitivity.

5. Model diagnosis. Determine the range of behaviors that are amenable to the model. Stress test

6. Model application and use. Test hypothesis and determine accuracy, failures, and limitations. Optimize

7. Adjustment and reiteration. Use data from previous steps to refine and retest the model.

Figure 8.5 ***Steps in modeling a biological system for the exposome.*** It will be necessary to build advanced computational models for the exposome. This figure illustrates a simple overview of the process. It begins with identifying the goals and objectives for the modeling exercise. What is the expectation from the model? This is followed by selecting the type of model to be used, which depends on the nature of the available data. The model will then be diagramed with various components and variables listed. Data are then entered into the model and the predictions are analyzed for consistency to published data and expert opinion. The model can be then used to test the effects of variable manipulation, which is not only similar to the second rung of Pearl's ladder, but also counterintuitive for counterfactual analysis that approximates the third rung. *Adapted from Voit, E. O., Qi, Z., Miller, G. W. (2008). Steps of modeling complex biological systems.* Pharmacopsychiatry, *41, 578–584 (Voit, Qi, & Miller, 2008).*

pathways and complex human behavior would be nice, the likelihood of such an approach working is very low. Thus, it may be necessary to revert to our reductionistic ways and develop more feasible models that examine one specific type of exposure or endpoint in the hopes of future integration. It is rather difficult to assess the quality of the data that are currently is because we do not know what it is we are trying to do. Ideally, once the objectives and scope are identified then data standards can be developed to determine which data would be suitable for analysis.

The actual selection of the model(s) is critical. First, we must determine the essence of the complex exposome. For example, we can state that external factors alter human biology. This is extremely broad, but it establishes the relationship. We need a way of modeling the external factors and a way of modeling human biology. Our model could benefit from having correlative and explanatory components. If an increase in one particular external factor decreases function of a certain biological system, we do not necessarily need to know why. If we did, it could help identify similar contributors and allow us to determine what factors are influencing the accuracy of the predictions. Given that we do not have all of the necessary information available at the outset, a deterministic model will not be appropriate. A probabilistic or stochastic model is better at dealing with randomness and variability, and those are certainly going to be components of the exposome. There are a large number of variables, signals, and parameters that will need to be defined. Once a model is sketched out and populated with various types of equations, it needs to undergo thorough diagnostics to determine if it is internally consistent, sensitive to changes in input, and stable under such challenges. It must be determined that the model is providing data that are reasonable at a qualitative and quantitative level. The next step involves the actual use of the model that starts with validation using experimental data and then moves to untested scenarios. Based on those experimental results, the model can be manipulated and optimized. Thus, it is an iterative process with multiple checks and balances. Ultimately, the goal is to have a model that provides accurate predictions of outcomes based on novel inputs. For example, does a particular combination of chemicals and stressors increase or decrease the likelihood of contracting a particular disease? This process is not for the faint of heart. Voit stated when referring to the exposome, "This is not rocket science, this is harder than rocket science. Space exploration is based on linear models, the exposome is based on non-linear models, which are much more complicated." I used this quote in the original edition of this book and had not planned the use of the moonshot analogy to illustrate innovation in Chapter 6, Innovation and the Exposome, as a means of reinforcing his point, but regardless "... harder than rocket science." Thanks, Eberhard.

8.10 Tipping the scales

The idea of small-world networks was introduced in the last chapter. These networks are defined by nonrandom organizations that are characterized by power laws (relationships where the change in one parameter confers a change in the second that varies as a power of the other). This concept is wonderfully illustrated by the fractal biology of nature in which structural patterns are repeated on progressively smaller scales (Mandelbrot, 1983). The geometric characteristics of a tree trunk, the branches, and leaves are defined by scaling exponents. The growth trajectories of animals converge on a single curve that appears to be universal (West, 2017). These universal rules that govern biological growth, aging, economic markets, and urban systems are also going to apply to the exposome. Our exposome can be viewed as a complex system both from the external forces outside of our body and the multiomic manifestations that result from those exposures. A reductionistic mindset will hinder our ability to make sense of the exposome and this serves as yet another reminder of the need to recruit investigators with expertise in complex systems or to train our workforce in how to study the environment and health at scale.

8.11 Predicting predictions

Predicting outcomes is a commonly stated goal of many health-related projects. We want to predict who will develop diabetes. We want to predict who will have a heart attack. We want to predict who will develop cancer. We certainly know that having an increased number of risk factors increases the likelihood of a person getting a disease, but in reality we are not that good at such predictions. Are we being unrealistic in thinking we can predict our exposome-regulated destiny with any level of precision? One option would be to stop trying. Rather than focusing on predicting outcomes that occur in the future, perhaps we should try to measure an individual's ability to respond to those exposome-encompassed external forces in the here and now. Which is harder to do, predict who will get cancer in the next 20 years or who will respond more favorably to an environmental insult in the next hour? It is very difficult for a cardiologist to predict who will die of a heart attack in the next decade, but she or he can perform a stress test on a patient and know immediately how the patient responds to the challenge of a bout of exercise (external stress is the work performed) and assess the patient's robustness or fragility with a high degree of precision. We can try to predict who will perform poorly in such a test, but we do not stop there. We actually test them. We do not rely on predictions. In the context of the exposome, we want to know how people respond to their past and current environmental exposures. We want to know what the cumulative biological response is and use that as a measure of fragility or vulnerability. Moreover, we can determine if a high chemical body burden (perhaps reflected in an exposome risk score, see Chapter 10, The Exposome in the Future) increases one's vulnerability to pathogens or other stressors. Thus we can fantasize about predicting our health futures, but the more realistic approach may be to quantify how our exposome is exerting deleterious effects on our health in the present, and takes steps to reduce those exposures in the here and now.

8.12 Ethics, biases, and data integrity

We need to come to grips with computers running larger parts of our lives. Hal in *2001: A Space Odyssey*, or Big Blue playing chess or winning on Jeopardy may not seem threatening, but when such systems are diagnosing our illnesses, deciding our medical care, or determining our retirement allocation, we may start paying more attention. I believe it is incumbent upon the users of the systems that impact the consumer to demonstrate how they solve less complex problems in ways that are logical and scalable. The ability of these systems to make obvious decisions in a way that is verifiable should be demonstrated to the people who will be affected by it. The decisions will be too important to ask the public to trust the organizations that are using that information to influence our lives. The decisions made by these systems for complex problems should be evaluated through peer review and expert panels, but the ultimate affected user (e.g., taxpayers) should have access to underlying data and systems in a manner similar to that of credit reporting systems.

One of the purposes of health-care systems is to provide the best medical care to those who need it. Unfortunately, in many parts of the world there are major disparities in who gets access to the best care and who gets access to a lesser quality of care. Although the exposome does have potential to incorporate information about environmental exposures into personalized medicine, the bigger impact, and the one that should be the focus of much of the work, will be on population level to prevent exposures. As we learn more about the environmental drivers of disease strategies can be developed to prevent detrimental exposures to thousands or millions of people. Thus, exposome research should be focused on strategies that improve health regardless of socioeconomic background, ethnicity, or education.

The populations to be studied should be representative of the population at large so that developed solutions are scalable. For example, if certain water contaminants are identified in a city, it should not only be the people who can afford under-the-sink water filters to benefit. The solutions should be implemented at the levels that prevent adverse effects in the entire community. For example, better solutions would be the distribution of clean water throughout the community while remediation efforts are underway, or installation of filtration systems at water mains that supply the entire community. If only wealthy people can mitigate the danger, then the scientific discoveries will only serve to exacerbate existing health disparities. In *The Ethical Algorithm*, Kearns and Roth not only provide examples of poorly designed systems but also provide guidelines for the use and development of data and algorithms (Kearns & Roth, 2020). Specifically, the authors describe how one could encode ethical principles into the algorithms that are being used to drive decisions that affect the general public or the individual patient. The authors emphasize the need for humans to be involved with the design and development of the algorithms. However, it is also critical that those humans have some level of domain knowledge. A computer scientist is human but may know nothing about the health condition being modeled, which illustrates the need for interdisciplinary teams to be involved in algorithm development. For additional examples of how automated systems and algorithms can wreak havoc on society, I recommend the book *Automating Inequality* by Virginia Eubanks.

8.13 On cloud 9

The origin of the term "on cloud 9" to denote a state of bliss is uncertain; some may think that cloud computing has the same ephemeral quality. Cloud computing likely does elicit bliss in some people (e.g., computationally demanding investigators, colleagues trying to share data across continents, shareholders of Amazon), but it is not as mysterious as the idiom. Cloud computing has been given a near angelic quality for its ability to solve all problems computational. But the reality is that it is merely a distribution system that consolidates the server farms we had previously within our own institutions. What it does do to a certain degree is to democratize computing. By providing access to some of the greatest computational machinery to users all over the world, one does not need to wait for their employer, whether it be a university, government, or corporation, to purchase the new high-performance computing system. Now it is possible to partner with cloud providers to rent time on their computers to solve problems. This is not terribly different to crowd sourcing of personal computers to solve various societal challenges. The primary difference is that this is profit driven.

Would it be possible for collaborative universities to create their own academic cloud to advance exposome and other environmental and biomedical research efforts? Something akin to the Azures, Google, and Amazons, but in a nonprofit model? For years, academic investigators have collaborated on computationally intense projects and shared computational power, but not at the scale that is needed for the future challenges. At this time, these efforts will likely be stifled by the start-up and projected maintenance costs, but it would be an interesting exercise to estimate how much money universities will pour into commercial cloud providers over the next decades. The cloud providers have been handing out cloud credits to lure academics, but their models are not appropriately discounted for the nonprofit academic sector (which ironically is responsible for training their future employees). The corporate goal is to get business (and universities) addicted to their products, move them away from home-based high-performance computing systems, and have them become more and more reliant on the cloud. There is no doubt that the publicly funded databases for genomics have been essential for advancement in that arena. EPA's investment in the Computational Toxicology Dashboard is welcomed but is unlikely to serve the needs of the biomedical and clinical research communities.

8.14 Next-generation training

The exposome needs computationally savvy investigators. Unfortunately, we cannot rely on being able to hire data scientists for our projects when there are extremely high-paying jobs in other sectors. I have been advocating an entirely different approach. We must identify investigators in environmental health science who have a propensity or desire to learn advanced data science approaches, work to develop them, and make it attractive for them to stay within the field. Providing outstanding training and skills to our trainees runs the risk that they may be lured away from other sectors, but this is a risk we must be willing to take. If trainees are provided with the skills to solve a major societal challenge (the exposome writ large), it may be possible to keep them engaged and challenged by those important problems. Given the need for a new generation of scientists who can apply modern data science techniques is needed to solve environmental health science problems, I have been striving to create this type of data science—savvy environmental health scientist in my various roles. Graduate students are already seeking additional training in data sciences by participating massively open online courses (MOOCs) and various workshops, but there is also a need to embed data science training into environmental health science curricula. It is not just for the students, as at the same time, faculty are grappling with increasingly large and complex datasets from climate science, epigenetics, the exposome, and multiomics, and they are in need of expertise to advance research efforts.

It is, therefore, essential that the next generation of environmental health scientists receives training in quantitative methods in order to fully exploit the value of the data being generated. Development of training programs that emphasizes competency in data sciences requires deep expertise in core quantitative disciplines and an appreciation of the scientific domains (here the environmental health sciences) to which techniques are to be applied. One key challenge is that much of data science comes from engineering and computer science, so the tools are not necessarily directly applicable to environmental health research. Fortunately, more and more environmental health researchers are gaining this expertise and can help bridge the disciplines. Areas that need to be covered include reproducibility and replicability, data harmonization, data visualization, causal modeling, cloud computing, mixed methods modeling, Bayesian probability, use of high-performance computing clusters, and computationally efficient algorithms. A combination of coursework, along with seminars, and workshops will be necessary to provide students and faculty the full array of quantitative training opportunities.

At Columbia University, we are integrating data science within our environmental health science training programs. The goal of the core curriculum is to provide training in the breadth of environmental health, including the exposome, the impacts of climate on health, and health across the life span built upon a foundation of advanced data science tools. Doctoral students will gain competency in several aspects of data science. This includes advanced coursework in data science as it relates to environmental health that combines classroom lectures with computational exercises. Students also participate in multiple workshops and bootcamps for specific skills. Courses currently offered at Columbia will provide the basis for this requirement, but students are encouraged to take training off-site through Google, Amazon, and Microsoft (we do have the geographical advantage of these companies that have substantial presences in New York City, but as these organizations grow more and more cities will have access to these resources). Students will be required to attend seminars that utilize data science tools and prepare brief summaries of the covered topics. Our students will also conduct a data science project within their small interdisciplinary training group. We also require students to attend key university events that are focused on data science to learn more about trends in the field and to develop their professional networks outside of environmental health science. To keep track of this multifaceted training, the students will maintain a data science portfolio, similar to how an art student maintains a portfolio of sketches (Fig. 8.6). The data science portfolio will chronicle their experience, skills, and gaps (i.e., areas of future growth). As part of comprehensive exams, students provide the exam committee their current data science portfolio that includes plans for any further training. The portfolio will provide an accurate representation of their data science skill set, and while not a certification it does provide detailed information on the skills the students have mastered.

8.15 Obstacles and opportunities

As data science methods, training, and careers evolve, the greatest challenge for the exposome field is keeping up with the advances in a meaningful way. We cannot rely on a strategy of picking up the crumbs from the more established fields. We must forge substantial collaborations with leading data science teams so that we can help drive their future innovation. Understanding the environmental forces that drive our health,

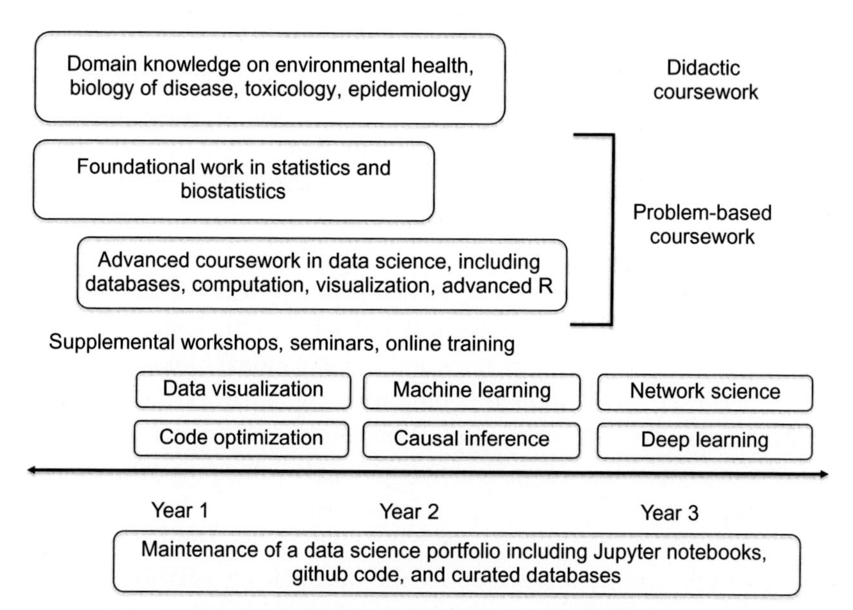

Figure 8.6 ***Data science training for the exposome.*** This figure shows an example of how data science skills could be embedded into a PhD curriculum of environmental epidemiology, environmental health sciences, toxicology, or other exposome-related doctoral training. Students take their typical didactic training, including biostatistics, but then have access to advanced coursework in data science that provides the fundamentals. Attendance at workshops, seminars, and online courses can supplement their skills needed for their dissertation work. All of the data science–related effort is compiled in a data science portfolio that illustrates the student's understanding or mastery of different aspects of data science. This allows for a tailored and nimble approach for each student.

from pandemics to urbanization, is a laudable goal and one that should be interesting enough for data scientists to want to tackle.

One of the great opportunities for the exposome is the possibility of generating interactive maps based upon detailed and multiplexed exposure and biological data, and this level of integration will require sophisticated analytical tools. Groups that study weather patterns and atmospheric modeling are dealing with similarly complex data, and collaborations with these groups represent an excellent opportunity. Over the past few decades, the ability to accurately assess weather patterns has grown tremendously. While weather patterns may become an important component of the exposome, it may be the techniques borrowed from that field that may prove to be the most useful. For example, the U.S. Department of Energy's Oak Ridge National Laboratories and the National Aeronautics and Space Administration have extremely powerful computer systems to process spatially related data. Similar to what was done under the Human Genome Project, government laboratories across the world will need to be brought into the fold. Successes in other areas of biomedical sciences can also have a positive impact on the exposome. The integration of exposome-level data within the medical field is a prime goal. Medical centers have gone to great lengths to systematize their electronic medical records (EMR). If we are able to generate exposure data for individuals, it will be important to make sure that it is amenable to the EMR systems.

If data science, including systems and computational biology, network science, and data visualization are going to become part of the exposome foundation, the field of environmental health sciences must integrate this type of training into our educational platforms. It will continue to be necessary to rely on specialists in these areas to some degree, but the investigators interested in the exposome will need to be familiar with the language, tools, and strategies used by the quantitative specialists (quants). This may be one of the more significant obstacles for the exposome in that environmental health scientists are already embracing the omic-type technologies, spatial statistics, and advanced epidemiological analyses, but the field is lacking in deep expertise in advanced data science methods such as machine learning/artificial intelligence, network science, and data visualization. Yet, these are not insurmountable obstacles but instead represent challenges to be overcome as the field matures.

8.16 Discussion questions

1. What should be the data-related competencies for a student pursuing exposome research? Statistics? Bioinformatics? Coding? Data visualization? Computational modeling? Are these competencies inherently different than what is needed for the biomedical sciences in general?
2. What roles and responsibilities do corporations have in assessing the exposome or in the application of the information derived from exposome research?
3. What steps need to be taken to protect the general public from the misuse of information? If detailed GIS-level information on environmental exposures was freely available, would it impact where you

choose to live? Would it affect property values? Would it affect your behavior, e.g. drive slower, eat less fast-food?

References

Costello, S., Cockburn, M., Bronstein, J., Zhang, X., & Ritz, B. (2009). Parkinson's disease and residential exposure to maneb and paraquat from agricultural applications in the central valley of California. *American Journal of Epidemiology, 169*(8), 919–926.

Forscher, B. K. (1963). Chaos in the Brickyard. *Science, 142*(3590), 339. Available from https://doi.org/10.1126/science.142.3590.339.

Juarez, P. D., Hood, D. B., Rogers, G. L., et al. (2017). A novel approach to analyzing lung cancer mortality disparities: Using the exposome and a graph-theoretical toolchain. *Environmental Disease, 2*(2), 33–44.

Kalia, V., Walker, D. I., Krasnodemski, K. M., Jones, D. P., Miller, G. W., & Kioumourtzoglou, M. -A. (2020). Unsupervised dimensionality reduction for exposome research. *Current Opinion in Environmental Science and Health*, in press.

Kearns, M., & Roth, A. (2020). *The ethical algorithm*. New York: Oxford University Press.

Krewski, D., Andersen, M. E., Tyshenko, M. G., et al. (2020). Toxicity testing in the 21st century: progress in the past decade and future perspectives. *Archives of Toxicology, 94* (1), 1–58.

Mandelbrot, B. B. (1983). *The fractal geometry of nature*. New York: W.H. Freeman and Company.

Pearl, J., & Mackenzie, D. (2018). *The book of why: The new science of cause and effect*. New York: Basic Books.

Schwartz, J. D., Wang, Y., Kloog, I., Yitshak-Sade, M., Dominici, F., & Zanobetti, A. (2018). Estimating the effects of $PM_{2.5}$ on life expectancy using causal modeling methods. *Environmental Health Perspectives, 126*(12), 127002.

Schenk, D. (2007). *Data smog: Surviving the information glut*. New York: Harper Collins Publishing.

Shea, D. O. (2007). *The Poincaré conjecture: In search of the shape of the universe*. New York: Walker Publishing Company.

Voit, E. O., Qi, Z., & Miller, G. W. (2008). Steps of modeling complex biological systems. *Pharmacopsychiatry, 41*, 578–584.

West, G. (2017). *Scale: The universal laws of growth, innovation, sustainability, and the pace of life in organisms, cities, economies, and companies*. New York: Penguin Press.

Wing, J. M. (2006). Computational thinking. *Communications of the Association of Computing Machinery, 49*(3), 33.

Further reading

Barabási, A.-L., & Pósfai, M. (2016). *Network science*. Cambridge: Cambridge University Press, Ethical algorithm.

Börner, K., Bueckle, A., & Ginda, M. (2019). Data visualization literacy: Definitions, conceptual frameworks, exercises, and assessments. *Proceedings of the National Academy of Sciences of the United States of America, 116*(6), 1857–1864.

Eubanks, V. (2017). *Automating inequality: How high-tech tools profile, police, and punish the poor*. New York: St. Martin's Press.

Newman, M. E. (2010). *Networks: An introduction*. Oxford, UK: Oxford University Press.

Voit, E. O. (2017). *A first course in systems biology* (2nd ed.). Garland Science.

Wickham, H., & Groelmund, G. (2017). *R for data science: import, tidy, transform, visualize, and model data*. Sebastopol, CA: O'Reilly Media, Inc.

CHAPTER NINE

The exposome in the community

In the aftermath of Washington, D.C., I knew that something like Flint was inevitable.

Marc Edwards, Virginia Tech

9.1 Community matters

Environmental health tragedies, such as the water crisis that occurred in Flint, Michigan, reveal breakdowns in our governmental structures and infrastructures. Communities know that something is amiss, but it can be exceedingly difficult for government officials to accept that their systems are not working. Often it takes the outcry of the community, in combination with concerned and knowledgeable scientists, to reveal the problem. This was illustrated with the crisis in Flint (and similarly in Washington, D.C. years earlier). What worries me, though, is that there are numerous instances of unintended exposures that do not rise to the level of what was seen in Flint, are not detected by the community or government, and are summarily ignored. Exposome research should help close this gap as it promotes widespread surveillance of the environment, including untargeted analyses that can detect unknown contaminants. A more systematic way of evaluating the environmental impacts on health should allow the examination of the health impacts of our environment at national and international scales. This will require engagement with our communities, inclusion of citizen scientists, and a democratization of exposome data and knowledge.

As noted in the preceding chapters, the challenges facing data analysis for the exposome are daunting, some may say impossible. I ask the reader to set the thoughts of impossibility aside for a moment. There is one aspect of the exposome that is not complex or fraught with challenges and scientific danger. It is my view that the exposome is a superb educational vehicle to promote the importance of the environment in health

The Exposome.
DOI: https://doi.org/10.1016/B978-0-12-814079-6.00009-2

and disease. The exposome is an accessible mental model that can be used by students and citizens alike. Each day we are bombarded by a dizzying amount of exposures and influences from our environment. We are aware of many of these exposures through our senses and intuition, and also through information coming from news outlets, social networks, advertisers, and our colleagues and family members, but we generally lack a framework to organize all of the information.

Recall that human intuition fails miserably at dealing with complex data. Most people cannot keep track of the various controllable environmental factors they face every day. Keeping track of a diet is difficult, but possible, yet remembering to not exercise during rush hour or near crowded highways, to avoid the stress of overscheduling, to avoid mixing certain household chemicals, to thoroughly wash fruits and vegetables, and to take any necessary prescriptions can be exhausting. Most people throw up their hands in defeat and just focus on what they perceive to be the most important. Unfortunately, their perceptions—shaped by the often well-meaning, but often misinformed media, friends, and family—are can be shaky from a scientific perspective. The exposome, especially charted as the modifiable version shown later, is an excellent way of organizing these multiple influences in a way that brings clarity to the morass of information. The development of an expanded, personalized, and easy-to-understand version of this could become an interactive tool to be used with one's physician, as a guidepost for home eating habits and as a blueprint for overall health.

9.2 Sensory and technical overload

The dizzying amount of information to which we are subjected is only increasing. Coming from multiple directions and sources, these data can blur our thinking and hinder our ability to make sound decisions. In business circles the term "systems thinking" has been proposed as the future of decision-making. I have never been a fan of the term "systems thinking" which may be a bit hypocritical given my admiration of the field of systems biology. I believe that thinking is a deeply complicated process. All thinking should be conducted at a systems level. Indeed, I like to define systems thinking as merely thinking. Reacting without thinking is by definition not thinking. That said, the concept of systems thinking is the type of true contemplation and analysis of downstream effects that is needed to understand our health, manage our environment, and make good decisions.

The foundation of systems thinking is to consider implications and impacts of actions beyond the first order. That is, most people have a short-term view in their decision-making process and do not consider second- and third-order consequences. I will use a rather apropos example here. Consider cigarette smoking. A smoker knows that lighting up a cigarette will bring them pleasure. Even if they do not know how the nicotine activates nicotinic receptors that lead to the release of dopamine in the reward systems of their brains, the user can appreciate the pleasurable sensation. Unless they live under a proverbial rock, they also know that smoking has adverse health consequences. Smoking increases the risk of throat, neck, and lung cancers. It impairs breathing capacity. It makes clothes smell like smoke. Those are the primary effects. Over the past few decades, we have become much more aware of the second- and third-order consequences. The most obvious example is the aptly named second-hand smoke. When a person smokes a cigarette, the lit end releases tobacco smoke into the air as does the smoker's exhalation of the inhaled smoke. Thus, the smokers who assume the risk of their habit on their own health introduce a second-order effect on the people and pets in close enough proximity to come in contact with the second-hand smoke.

Unfortunately, the downstream effects of smoking continue. Third-hand smoke represents a less intuitive event. If you have ever entered the home of a smoker, it smells like smoke even if nobody is actively smoking. This means that chemical residues from the smoke must reside in the house. Odors are chemical in nature. Therefore, if the chemicals that are responsible for the odor of burnt tobacco remain in the house, it is certain that other chemicals in cigarette smoke are also going to remain in the home. Cadmium, polycyclic hydrocarbons, and many carcinogens will be in the carpets, upholstery, curtains, and even the paint on the walls. Much of this can also be transferred into dust. Thus, if you purchased the home of a smoker, you will be exposed to the chemicals the smoker deposited all over the house. Our actions have downstream consequences, and few put much thought into this. However, the downstream consequences of our everyday activities can have lasting impact on the individual and those around us.

I find it instructive to consider one's habits and activities in this mindset. It is not as if you have to conduct this analysis every time you engage in any activity. Just spend some time thinking about your daily activities and how they impact yours and your family's health in the short and long term. Once you map out these major drivers of your health, you can make decisions about them in a holistic way. New data that come your way can then be plotted onto your exposome graph. It should fall within one of the major domains you have already identified, providing a way to filter the new information. Individuals may need input from professionals or experts. This could be a physician, nurse, dietician, or public health professional. Although physicians generally do not have a great deal of time, they can react to a diagram that maps out your current health state.

The illustration in Fig. 9.1 shows some ways that the Example family can improve their exposome; however, there are also behaviors that can worsen their situation. If one of the members of the Example family were to start smoking in their home, all of their family members and Sparky and Whiskers would be exposed to the second-hand smoke and all of the associated toxicants. There is substantial evidence that family members of smokers who smoke inside the house have an increased risk of pulmonary disease. Veterinarian pathologists will tell you that it is very obvious to identify pets whose owners are smokers. Postmortem, the lungs of dogs and cats look very much like those of a smoker. It is safe to say that the

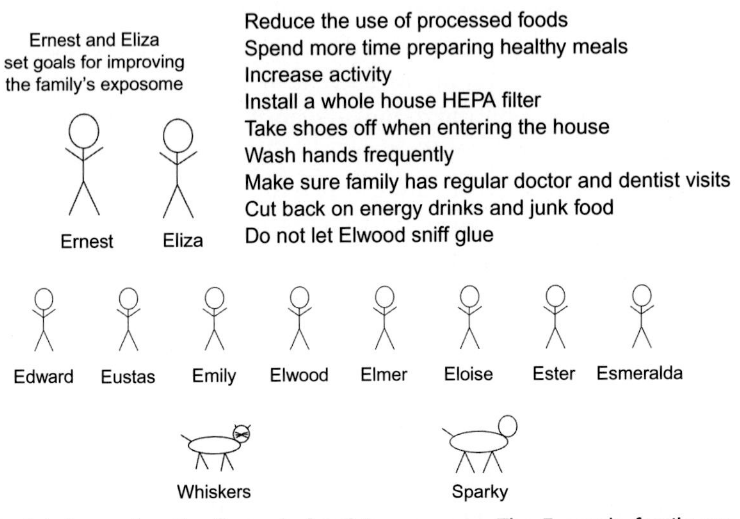

Figure 9.1 ***Improving the Example family's exposome.*** The Example family can take several steps to improve their exposome. Some steps require planning, some take time, and some may require some financial investment. However, most of the improvements are based upon improving behaviors.

average smoker is not considering the impact that their smoking will have on their own pets, or even the young family that moves into their house when they sell it in the next year.

9.3 The modifiable exposome

There are major national and international policy implications for the exposome, but some of the most rapid opportunities occur at the individual, family, and community level. Here we will examine how the exposome can be modified at a local level. The components of the exposome display varying degrees of impact on one's health. Cigarette smoking is much more deleterious than the injestion of the occasional hot dog. Some components are very much under our control (smoking and physical activity), while some are influenced more at a population or societal level (air pollution and healthcare access). Fig. 9.2 illustrates the relative

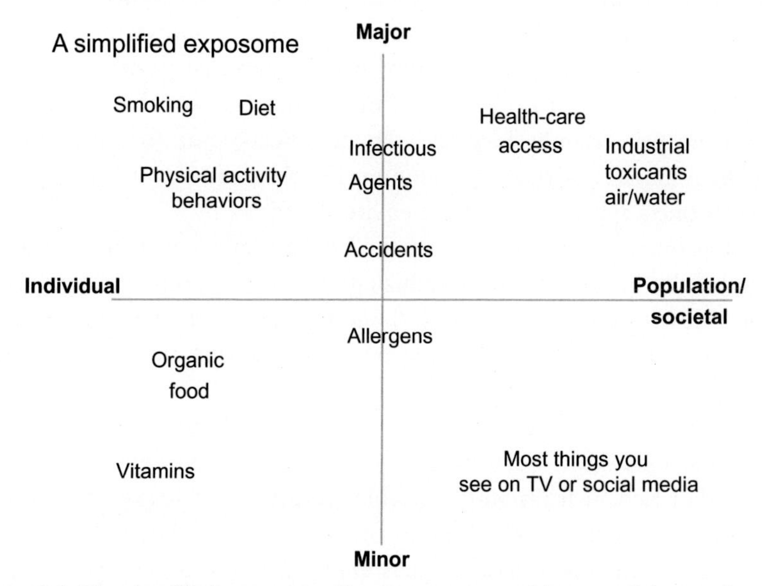

Figure 9.2 ***The simplified exposome.*** The exposome provides a useful mental model to organize the multiple influences on our health. By using an intentionally broad definition of environment, nearly all of the external factors that influence one's health are included in the exposome. It is then possible to plot these various factors along axes that include the importance to health (minor to major) and the ability to control said factors at the individual, community, or societal/population levels.

contribution based on these two scales. The y-axis shows the relative degrees of influence, while the x-axis displays the level of control from individual to population. Items in the upper half of the diagram have major influences on health and we should be taking steps to minimize their negative impact or to enhance their benefits. Items on the bottom can certainly influence health; they just take a lower priority. The items on the left are under the control of the individual or the community to some degree, while those on the right are more on a population or societal level. The top right quadrant contains the major issues that must involve governments and healthcare systems. Improving healthcare access, reducing air pollution, and improving water supplies all require some intervention on a large scale that involves populations, governments, and regulations. Of course the individual has a role in voicing concerns to their elected officials, but the ultimate solutions come from the larger organizations. Climate change was left out of this graph. This was not intended to minimize its importance, but because the timeframe of the graph is relatively short. The idea is to explore how people can modify their exposome in a positive way on the scale of days, months, and eventually years. This may be temporally naïve, but it allows the reader to focus on the more immediate future from a personal health standpoint. The proper care and maintenance of our planet is essential and the exposome concept certainly includes climate effects, but focusing on the immediate impacts of our exposome on health is easier to consider and it enhances one's appreciation of the environment.

Eating organic foods is listed in the bottom left. This is not meant to downplay the importance of minimizing chemical exposures but to take a broader view of the topic. Overall, the judicious use of chemicals increases crop production and keeps disease and pest damage in check. From a global perspective, organic produce is a luxury. Certainly, if one can afford it, it is best to minimize one's exposures to pesticides and other agricultural chemicals, but eating organic is not as important as eating a healthy diet rich in fruits and vegetables, and for that reason it is viewed as minor. The benefits of eating an organic diet may be as much about the quality of the food itself and less about the lack of pesticide residue. A move toward lower pesticide use across the entire agricultural sector would be of greater benefit than the expansion of the organic industry (although the latter may be helping drive the former). From a food quality perspective, processing and packaging may be a better way to evaluate the potential exposure to chemicals. It is my view that eating fewer processed foods is a higher priority than eating organic (although doing both would be ideal). Processed foods contain many more chemicals and fewer nutrients. More packaging means more plastics, phthalates, and preservatives. The modern diet with its focus on convenience has increased the delivery of chemicals and decreases the quality of nutrients. This must be reversed. It is tempting to opt for the inexpensive convenient food, but it comes at a price: higher caloric density, higher fat content, higher sodium content, and more unnecessary synthetic chemicals. Fewer people want to spend their time in kitchens, and that is not good for the health of our society. Foods prepared at home with moderately healthy practices are superior in nutritional quality and lower in chemical residues. As adults, one of the most important health lessons we can provide children is to teach them how to prepare healthy foods. There are too many teenagers and college students whose cooking skills limit them to ramen noodles and fast food. This leads to terrible habits that can persist into adulthood.

Accidents straddle the line between individual and population control because there are individual steps (seatbelts, protective eyewear, safe practices, and common sense) and societal steps (OSHA regulations, highway safety interventions, and workplace practices) that can be taken. Notably, according to the Centers for Disease Control and Prevention (www.cdc.gov/injury), unintentional injury is the leading cause of death in the U.S. for all ages between 1 and 44! While we tend to think about the importance of heart disease and cancer, of the risk factors that do start at an early age, it is accidents that are most dangerous for most people until middle age. Although we rely on government to provide regulations, we should heed the advice of safety professionals. Alluding back to Chapter 2, Genes, Genomes, and Genomics: Advances and Limitations, when the late Francis Crick was asked about the key to a long healthy life he stated, "Always use the handrail!" Simple advice from an extraordinary scientist. Instead of referencing DNA, he was talking about his exposome (he just did not know it at that time).

Allergens are listed below the *x*-axis because for the majority of people allergens have a minor effect. Of course for individuals with severe allergies, the allergens may pose a much greater effect, and may even be one of the most significant drivers of health. Obviously, if such a graph were personalized and accounted for the individual susceptibilities and vulnerabilities, the layout could look much different. The ultimate goal of this exercise is to develop a personalized assessment and plan for optimizing one's exposome.

It is important to remember that each axis is a continuum. We have some level of control over our exposure to toxicants in the air or water—for example, maintaining vehicles, not exercising near highways, using home air purifiers, HEPA-filtered vacuum cleaners, and using water purifiers depending on water source. However, the major drivers (pun intended) are at a higher level. The number of vehicles on the road, exhaust regulations, and industrial pollution controls must be addressed at the regulatory and societal level. It is unfortunate that important targets for vehicle emission standards can fall victim to political power. That said, we do exert some level of control over our own transportation habits. While cigarette smoking is on the left side, we know that community and societal issues are also at play. From a societal standpoint, cigarettes should not be available to children and teenagers. We know that the sooner people start smoking, the more likely they are to become addicted. While raising the drinking age to 21 in the United States was a slow process and required significant governmental intervention, it did occur and has had a positive impact on the rates of accidents and injury. Had the U.S. tobacco settlement included a similar raising of the legal smoking age to 21 (phased in over several years at the same time that they distributed funds for various programs), it would have done more to improve the health of the U.S. than all of the billions of dollars that were poured into state coffers (even with a predictable prohibition-like black market). In the U.S. we were greeted with some unexpected news on this front in 2019 when the national age for the purchase of tobacco products was raised to 21 and this included electronic cigarettes. Although it will take some time for the effects to be realized, this legislation will likely have a major impact on public health by reducing tobacco and nicotine use in young adults. Unfortunately, the e-cigarette industry had a substantial head start and through aggressive marketing has produced thousands of addicted future customers.

The "most things on TV" description in Fig. 9.2 is only partially tongue in cheek and should also include social media. Most of the news stories related to health are sensationalist. They often focus on obscure, but newsworthy exposures. Decades ago I recall seeing a television news piece on the Marlboro man that included a critical assessment of the tobacco industry. This was when information regarding the cover-up of adverse effects by the industry was becoming public. Today, it is rare to find anything on television that is remotely educational concerning environmental factors in health. An unfortunate example about a lack of societal control is the use of e-cigarettes and vaping mentioned previously. On the surface, shifting from high-temperature burning of tobacco to nicotine delivery systems that contain few contaminants is a good idea. It is reasonable to conclude that vaping is safer than traditional cigarette smoking, but it is not without risks; the metal coils and flavorings can expose the user to a host of chemicals and the dose of the addictive nicotine can be higher than in cigarettes. More important, the lack of regulation created a public health travesty by allowing the industry to create a new generation of nicotine addicts. Adult cigarette smokers should have access to tools to reduce their exposure to tobacco smoke. Children and teenagers should not have legal access to tobacco or nicotine delivery systems and the new legislation in the U.S. is welcome news. Similar consideration should be given to "energy drinks" with high levels of caffeine. Moderate doses of caffeine, like those found in tea, do not appear to have adverse health effects, but the higher doses found in strong coffee and energy drinks should not be consumed by children and teenagers. In high school, I conducted a science fair project that examined the effects of caffeine on the development of turkeys [this was conducted in a scientific setting at the US Department of Agriculture (USDA)]. Suffice it to say, caffeinated coffee stunted growth, while decaffeinated coffee merely stained the turkeys' feathers (they are messy drinkers; Miller, unpublished observations). I was fortunate to learn this lesson over 30 years ago, but now I witness increased access to highly-caffeinated beverages that are undoubtedly marketed to children.

Infectious agents are listed very close to the top of the graph in Fig. 9.2. Of the items in our environment, infectious agents have the potential to have the biggest immediate impact on health. While many factors involved in the spread and contraction of disease are outside our control, the individual can be sure to obtain recommended vaccines, including seasonal flu vaccines, minimize contact with sick people, and follow appropriate hygiene practices. Not only has the misinformation about vaccines caused outbreaks of measles and other diseases, but I posit that it has also suppressed uptake of season influenza vaccines. This creates considerable challenges to governments and healthcare systems that are trying to protect the public from these diseases. Communities must work to encourage these important activities, and professionals must continually work to counter the pseudoscientific claims that pervade the public thinking. The COVID-19 pandemic has illustrated the need for an improved understanding of the external threats that infectious agents pose. Nearly every action we can take to improve our health is a result of a change in our exposome. Let us explore how we can view these changes through the exposome lens.

Imagine if you were to visit your general practitioner and share this graph (see Fig. 9.3). Each point represents a health factor that you and the physician deem to be important for your health. It is remarkable that all of these are part of the exposome. Even though you may note a particular family risk factor, all of the behaviors going forward are components of the exposome. Each factor is mapped onto the diagram to indicate its relative importance and the evaluation of where you are in these measures. For example, your total cholesterol may be high. It shows in the upper left quadrant as being important to your health but being in need of improvement.

Now that each of these factors is mapped, we can color code them to reflect their dynamic state. If you see an improvement in your cholesterol status, it would shift to green and move to the right (see Fig. 9.4). Thus it becomes easy to see how alterations in your exposome (as reflected in

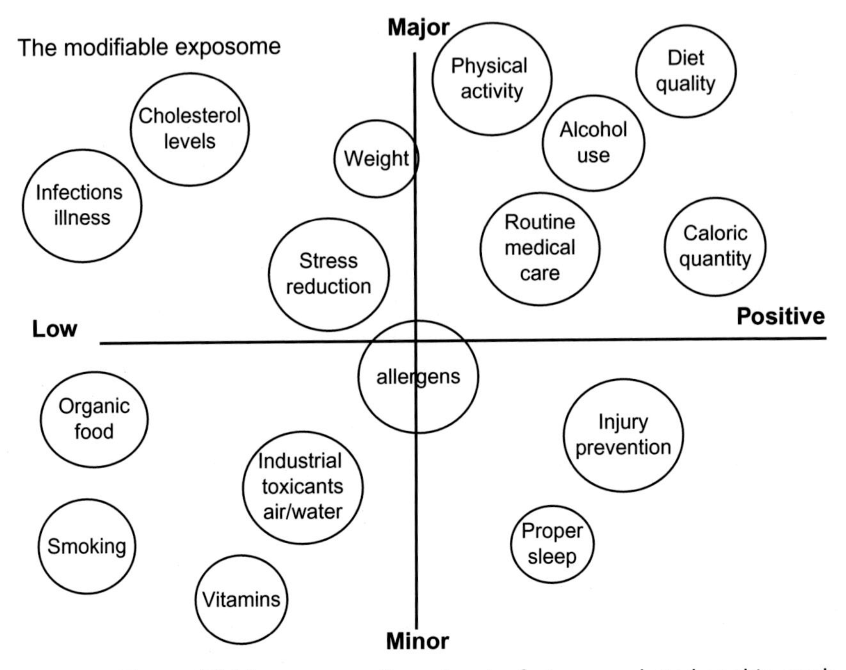

Figure 9.3 ***The modifiable exposome.*** Several major factors are plotted on this graph. Your graph may appear different depending on your background and activities. It is relatively easy to see how changes in each of these items can lead to an improvement in health.

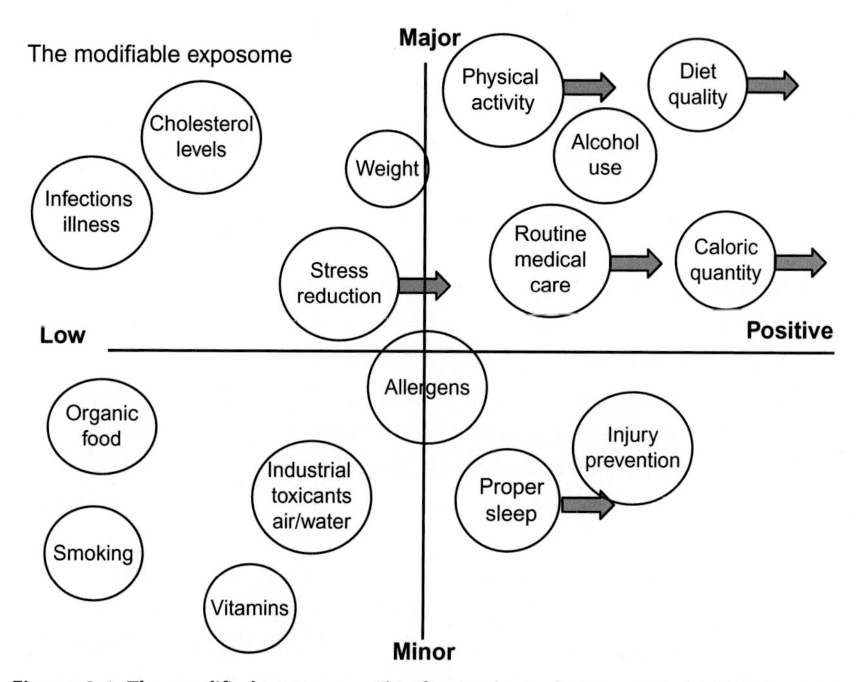

Figure 9.4 ***The modified exposome.*** This figure shows how one could plot the positive changes that have been made to improve one's exposome. Improved dietary habits, increased activity, and avoidance of negative behaviors can cause fairly dramatic changes in one's exposome and health. Imagine reviewing this plot on a weekly or monthly basis to evaluate overall health status.

lifestyle changes) shift your health status into a positive direction. But it is also mapping more than the singular item that you may be considering. Body weight or body mass index, diet, and exercise are interconnected and could be merged into a single unit. The vector could change based on changes in of the three components. Indeed, with a simple app, it should be possible to track these health factors over time—even provide a historical perspective. This becomes an exercise in data visualization as was discussed in the previous chapter. Making data understandable makes them more actionable. When people consider relocating, they do not generally think in these terms. They will evaluate job potential or salary, but not how that new environment will impact their health. Will it become easier or hard to exercise? Will pollen levels or particulate matter levels in the air change? Does the work environment promote health activities? Does the city provide reasonable public transportation? Are their positive commuting options? Will the environment be more or less healthy? Will there be new exposures? Will the quality of their diet increase or decrease?

The previous discussion was very much focused on the individual, but what happens when a person does not have control of many of these factors? What if the only apartment you can afford is located near a train station that has a power plant nearby? The increased noise and pollution become unavoidable. One may be able to mitigate the impact of noise on sleep by using earplugs or sound machines, but what about the pollution? Will indoor air purifiers reduce the in-room air pollutants? Can the person afford such luxuries, especially considering the cost of replacement HEPA filters? This is where policy comes into play. Governments have the authority to alter train traffic, to require antipollution devices on power plants, and to rezone areas to not allow housing near such sites. The challenge is getting the government to make thoughtful decisions to support their citizens.

Let us look at the modifiable exposome at a population scale (Fig. 9.3). For example, consider a community that is located near a factory. The factory provides jobs for many people in the town, but it also may produce a high level of emissions. Most people in the town cannot remember when the factory was not the major driver of the town's economy. Let us place our Example family in this town. Eliza's mother worked here for four years after college, but before she went back to graduate school for nuclear physics. The Example family lived here for six years. It was discovered that the factory was exceeding its emissions by 200 percent. This was not discovered by the regulatory body for two years. So the Example family was exposed to elevated levels of emissions for two years. Could this have contributed to the increased asthma and allergy symptoms the family had observed? Will there by long-lasting consequences of these exposures?

Ideally, we prevent these exposures before they occur, but it is almost impossible to prevent all of them. Not to be fatalistic, but governments do not tend to work on a timescale that is terribly useful for the individual's health. Changes in government policy can take years and implementation can add onto that time. However, these changes are essential for those who follow us, whether they be younger siblings or future generations. Therefore, as we continue to improve the conditions in our community and world, we must also take steps that to protect us from those issues that are immediately impacting us. One of the most important concepts to health is that of resilience.

9.4 Resilience and the exposome

As the conditions around us are constantly changing in predictable and unpredictable ways, we must prepare ourselves to resist or adapt to these conditions. There is a large body of health literature on the concept of resilience. We understand what it means to bounce back from a challenging event or life circumstance, but what does resilience mean in an exposome context? Can we develop resilience to exposures? It is obvious that when we do not get enough sleep, do not eat well, and engage in unhealthy activities, we increase the likelihood of getting sick. That is an example of decreased resilience or resistance to infection. It is reasonable to assume that this simple analogy is scalable. How does one become more resilient to exposome-based challenges?

Table 9.1 provides steps one can take to improve one's exposome. Most of these steps are obvious and do not require the exposome concept, but there is not generally a single source for this information. Rarely would a physician address each of these areas. A dietician or nutritionist would only address some. A mental health professional would address anxiety, stress, and cognition, but likely not mention diet. A fitness coach would address one's physical health and may address nutrition, but would not discuss safety practices, water quality, or cognitive activity. This leaves the individual as the only likely person capable of considering and integrating all of these external forces.

The exposome model is also a useful way to address health disparities. Variations in the factors that contribute to the exposome are far more influential than the genetic factors when one looks at various populations within a particular region. Poorer areas tend to be in closer proximity to hazardous waste sites and farther from parks and healthcare facilities. Juarez and his research group are using an exposome-based approach to evaluate such disparities and how they may impact health (Juarez et al., 2017). Layering rich exposome-type data on top of census and other socioeconomic data can help one to reveal trends among various exposures. The addition of comprehensive biological data could bolster these analyses, but it will be necessary to maintain a strong geospatial component to exposome data collection to facilitate the layering of these data. Eliminating health disparities is ultimately about modifying the exposome in the vulnerable and underserved populations. Genomics and precision medicine do not address the public health structures that help one to improve a person's exposome.

Table 9.1 Creating your enhanced exposome.

Intake
Reduce caloric intake (unless you are already at an ideal weight and body mass index and will lose weight by reducing intake)
Minimize intake of processed food. Recall that more processing and more packaging means low nutrient content and higher chemical content, including plasticizers and preservatives
Eat fresh vegetables and fruits. First priority—wash them. Second priority, buy organic if it does not impact your ability to purchase large quantities of vegetables. If it does, fresh nonorganic produce is better than processed food
Minimize alcohol usage (<1 drink per day preferably wine). Consuming large amounts of alcohol is associated with many adverse health consequences
Avoid the use of tobacco products (only use e-cigarettes if it is for smoking cessation)
Activity
Reduce injury risk—occupational setting, falls, gunshots, knife wounds, high-risk activities—wear proper protective equipment, get proper training for high-risk activities
Engage cognitive circuits—be sure to challenge yourself mentally throughout the day
Maintain relationships—avoid isolation—develop structures to avoid loneliness
Develop strategies to minimize stress—yoga, meditation, exercise, quite time. The most important part is identifying strategies that reduce your stress
Practice safe driving and transportation (safety belts, speed limits, and helmets)
Disengage from technology on a regular basis—cell phones computers have become fixtures in our modern lives. It can be hard to function in their absence, but it is necessary to disconnect. Identify spans of time where you are not looking or responding
Be active. Engage in cardiovascular and strength conditioning on a regular basis. Use the stairs, park farther away from the store, take vigorous walks, go to the gym, and participate in recreational sports

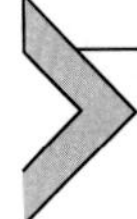

9.5 Engineering communities to optimize the exposome

Community members live within a physical environment that includes roads, parks, streets, houses, schools, and other buildings. This built environment has a major impact on our well-being and exposome. The built environment tends to be relegated to engineers and urban planners. Historically, health was not a major priority in urban planning. This is unfortunate because improving the built environment is one of the most effective ways of improving one's exposome. However, the built environment is increasingly being recognized to be crucial in creating healthy and productive communities. The proximity of housing and sidewalks to roads and associated traffic-related pollution can be a major driver of exposures. The buildings in which we live and work can impact our air quality and influence exposures to volatile substances and airborne materials. The materials used for the construction of our work-related buildings and offices, apartments, houses, and furnishings can have significant effects on our exposures. Many chemicals are used to increase the lifespan of counter surfaces, carpeting, and paints, such as the wood preservatives formaldehyde or arsenic. The most obvious example is lead-based paint. Lead allowed paint to last longer. It prevented peeling and chipping and reduced the time interval between repainting. However, we now know that the lead in paint was a major source of lead exposure and harmful health consequences. The paint was the vehicle for a major public health crisis. This suggests that all of the materials in our built environment have the potential to pose a risk to our health. As noted previously, health is generally not a priority in these decisions (arguably, it is not even considered except for avoiding using components banned by industrial standards). We know that fires can be devastating, so minimizing the flammability of building materials has been a high priority. This led to the widespread use of flame retardants in all sorts of furnishings and clothes. We are now learning that these materials, fluorinated and brominated flame retardants, have adverse effects on health and they are ubiquitous and persist in our environment. Reducing the immediate risk of fire-related injury was the goal, but long-term consequences were not considered.

9.6 Behaviors, mandates, and nudges

The concept of nudge, which garnered widespread attention with the publication of *Nudge: Improving Decisions about Health, Wealth, and Happiness* by Richard Thaler and Cass Sunstein, is to gently guide people into positive behaviors by making it easier to make the right decision (Thaler & Sunstein, 2008; Sunstein, 2013). The decisions that people make regarding their environment are based on a combination of factors. First, the individual can know what is best for him or her and pursue that particular action. Second, the individual can know what is best for him or her and choose not to pursue that particular action. Third, the individual can be ignorant of what is best for them and may or may not pursue the appropriate activity. For each of the dozens of choices to be made in a single day, there will be various combinations of the aforementioned decisions. Often the poor choices we make in maintaining a healthy exposome are based upon our own laziness. We may well know what is best, but because such decisions can require more effort, money, and energy, we choose not to make them. The conservation of energy may appear to be a positive adaptation from a bioenergetic standpoint, but it leads to small negative consequences that can build up over time to weaken the individual. Thus, there is an advantage from a policy, regulatory, and organizational perspective to make good choices the default option. We see this sort of strategy used in many life settings. For example, the opt-out setup of a company 401k or retirement plan makes the default to save money. One must consciously choose not to participate in a retirement plan. When participation in such a program is opt-in, the participation rates are much lower.

Antismoking policies are designed in a similar manner. The default behavior in settings with such restrictions is not smoking. While the basis of such programs is often framed in the manner of reducing second-hand smoke, which is a good thing, it also forces behavioral changes in the smoker. With the majority of the world population being nonsmokers, antismoking policies are well accepted by the general public. The general population notices that there is less smoke in their presence and there is no downside. Things get trickier when we try to encourage behaviors that include the majority of people.

A recent example of this was the school lunch initiative in the U.S. to improve the quality of the food served in public schools. It made perfect sense. As noted previously, our children need to eat healthier food. If only healthy choices are made available, the children will have no choice but to eat healthier food (in general, the healthier choices have lower levels of preservatives and higher levels of nutrients). The USDA has control over the school lunch program, so instituting broad changes was easy. The result was a major pushback by children and school systems due to the swift and dramatic implementation (due to the children's unfamiliarity to the healthier choices and the impact on school system budgets). A more staggered implementation could have led to broader adoption and slowly shifted children's eating patterns and also muted the pushback from the organizations that were affected by these changes. The concept of nudge in the hands of a politician or a federal government can be interpreted or perceived as a shove, but that does not mean that prohealth policies should not be pursued. New York City's laws that restricted the size of sweetened beverages were initially received as an infringement on personal freedom. Although reduction of sugar intake is good for health, creating rules and laws that make swift and dramatic changes in behavior does not go over well. It should be noted that New York City's anti-tobacco laws faced similar resistance, but led to dramatic health improvements and served as a model for many other cities. Fortunately, consumers do adapt, and over time the new polices become part of daily life and support improved decision-making.

Building upon his work in *Nudge*, Cass Sunstein highlighted his accomplishments regarding the development of easy-to-understand government policies during his term in a subsequent book *Simpler*, although he did not acknowledge that the Affordable Care Act, which was introduced while he was still overseeing the release of regulations as director of the White House Office of Information and Regulatory Affairs, was an example of precisely the opposite of what he claimed to be doing. The lack of broad support for the legislation led to an onslaught of challenges that have been undermining a program that addressed a problem that the majority of the population agreed needed to be fixed. Perhaps I am naïve to think that competing political parties can agree on anything, but one party legislation always faces dramatic opposition when the other party regains power. Although it seems obvious, policies that involve a broader range of supporters are much more likely to be sustained even if they result in compromised goals.

9.7 The exposome in the classroom

If offered as a stand alone course or as a component of several courses, the exposome could impact students throughout their education, starting in elementary schools by providing an analogous construct to the genome allowing the nature—nurture comparison to be made in a systematic way. The exposome could also be included in biology and health classes. In high school, it would help one to frame the variables in our complex world to link earth science courses with those that address health. For college courses the exposome provides a key biological and health principle to nonscience majors. For science majors, it helps faculty connect the challenges students hear about with climate change, psychological stressors, and diet with their chemistry, biochemistry, and biology classes. Later, I will examine how the exposome can be integrated at multiple levels of education.

Concerning the field of public health, most schools have very little basic science in their curriculum. Basic concepts in biochemistry and physiology are often overlooked because it is not clear why students need this information. A course that examines the balance of genetic and environmental contributors to disease can provide an excellent introduction to many of the important scientific concepts in public health that make it easier to understand the various drivers of health. The exposome provides an organizational structure to understand the complex exposures that impact health, exposes students to cutting-edge science from multiple disciplines, and forces them to think in a holistic manner. It reminds them of the rapidity of scientific advancement and serves as a reminder to keep up with the literature. We experimented with a course called Genome, Exposome, and Health several years ago at Emory University. By providing information on both the genetic and environmental influences on health, the students could appreciate the balance between the two and recognize where their actions have the greatest effects.

While medical students learn about human physiology *ad nauseam*, they hear little about air pollution and atmospheric chemistry, environmental and occupational health, and physical activity. For the most part, the medical field has been dismissive of holistic medicine practitioners, such as homeopaths and chiropractors, based on the tenuous scientific foundations of their disciplines, but these practitioners try to treat the entire patient. Individuals in the general population yearn to be treated as a whole person, and these aforementioned practitioners give them what they want. Traditional (also referred to allopathic) medicine has missed this point to some degree. The average person does not want to be looked at as a renal system, a digestive system, and a cardiovascular system. They know that certain activities in one area of their life can impact another. Sending a person to four different specialists pretty much ensures that they no longer understand the causes and effects of their actions to their health. It is critical to distill the complicated medical minutia into structures the general medical consumer can understand and appreciate. Since the exposome encompasses all of the malleable components of our health, it can serve as an excellent rubric to understand health, and the behaviors and actions that improve health or contribute to disease.

When I initiated my work on the exposome, I was fearful that the general public would bristle at such an esoteric scientific concept, but I have found quite the opposite response. It has more been along the lines of "Well, duh, we know this; we have just been waiting for the scientists to figure it out." The vast majority of people want to understand the various factors that affect their health, but we need to work on our delivery of the information. With environmental health not being taught in most U.S. medical schools, we cannot currently rely on the medical field to deliver this message.

An example of a misunderstanding of the external factors that influence our health is the use of nutritional supplements and vitamins. A story in the monthly magazine *The Atlantic* by Paul Offit not only heaped praise upon Linus Pauling, one of the greatest chemists of all time (another one of those double Nobel winners, although one was for Peace so he has nothing on Sanger, Curie, and Bardeen), but also delivered equally biting scorn for his promotion of high doses of vitamins (Offit, 2013). Current research suggests that such vitamin regimens have no positive effect on health, but the general public has a view of "if vitamins are good then more must be better," and the health practitioner does little to counter this way of thinking. Most people cannot fathom that supposedly healthy vitamins could be worse than exposure to low levels of pesticides (I am not saying it is, just that it could be). Swallowing gram quantities of vitamins that obviously enter our body with ease may well be worse than exposure to pesticide drift. An old health adage is that once there is adequate caloric and nutrient uptake, removal of factors has a greater effect on health outcomes than adding them. Removing a deleterious influence typically trumps adding a positive influence. To put it another way, eliminating bad habits is likely to have a greater benefit than adding good habits, and it is possible that ingestion of megadoses of vitamins may be one of those bad habits. In addition, such vitamins and supplements may interfere with the proper absorption and metabolism of key nutrients and pharmaceuticals.

It is the proper management of one's exposome that is key. People put so much literal and figurative stock into the vitamin industry, but much of this is due to misperception and aggressive marketing. In the context of the exposome, the use of vitamins may be an important part of one's health regimen, but if the nutrient level in one's diet is already high (because they are eating a proper diet), it is more likely than not to be introducing a deleterious iatrogenic effect. A recent review showed no beneficial effects of vitamin supplements on major health outcomes (Khan et al., 2019). By looking at all of the sources of chemical exposure in one's life, the individual exposures can be placed in the appropriate biological context. In fact, removal of excessive nutrients and pesticides would be the best outcome. The limited oversight that the U.S. Food and Drug Administration has over the vitamin industry has led to massive consumption of these products that are not only a waste of money, but also a source of uncharacterized exposures. This is one of the areas in which precision medicine could have a great benefit. If we analyzed an individual's diet, exposures, etc. and determined that they would benefit from a nutritional supplement, then they could be recommended to take that specific supplement, but if their nutritional status was good there would be no need for the introduction of vitamins. It is important to note that the ingredients in vitamins are not inert. The fat soluble vitamins can accumulate in the body and raise to toxic levels. The trace metals contained in vitamins also pose a concern. There is evidence that the recommended daily allowance of manganese may have adverse consequences on the regulators of oxidative stress. That said, except for the daily multivitamin that provides the RDA levels of vitamins and minerals, the potential adverse consequences of megadoses of vitamins, at least the fat-soluble ones, likely outweigh the often dubious potential positive effects. A sound understanding of how the exposome influences health may allow the consumer to make more judicious decisions about health habits.

9.8 Building a healthy exposome for the community

Another area in which the exposome can help one frame health effects is the built environment. Urban planners know how to design a community that includes walking trails and parks. Architects know how to design buildings that have minimal impacts on the environment. They know how to utilize more environment-friendly building materials. They know how to make the use of mixed-use developments. It is unfortunate that governments do not do more to reward more intelligent community and housing developments. Leadership in Energy and Environmental Design (LEED) certification promises users that future cost savings will offset the increased building costs, but communities also benefit from this type of thoughtful construction. The private organization has been instrumental in getting more environment-friendly building to occur and more recently has introduced the concept of LEED communities that encourage more expansive activities, such as redevelopment of sites in need of remediation from past toxic waste exposures.

As more people move from rural areas into cities and metropolises, this will become increasingly important. LEED also has a less-appreciated effort on human health. The environmental quality category focuses on how the materials used impact air quality. This category includes issues around ventilation, acoustics, reducing off-gassing of building materials, and the use of more natural lighting. LEED emphasizes research that shows that improving these areas can boost worker productivity, but there is also evidence that they improve health. However, this is an area that needs more research. There are few rigorous biology-based studies that examine the impact of building materials on health. Currently, LEED certification is a wholly voluntary activity. There is no doubt that many of the steps in LEED construction have higher costs, but many of these costs are offset over a relatively short period of time in energy or other utility savings. For example, improved natural lighting can decrease electricity use, and low- or no-flow toilets can provide substantial savings on water usage. However, it is not easy to make the larger upfront investments required for these improvements, and this is where building codes or tax incentives can promote the construction of buildings that support human health rather than antagonize it.

9.9 Teaching the exposome concept outside of the classroom

Academic investigators, well trained in the art of dissemination among their colleagues, often fall short when it comes to explaining their research to the general public, and this statement may even be too generous. When one looks at recent stories regarding environmental factors in health, one sees more sensation than information. How valuable is it to dribble out information on environmental health exposures to the general public? Given the current state of environmental health knowledge in the public, one could argue that it is a terrible approach. The public will remember the sensational story about one type of chemical and then irrationally never use nonstick cookware (but instead continue to use oil, suboptimal for health, on a traditional pan for decades). The public is fickle. A particular food is associated with a decreased risk of cancer and it becomes the trend du jour. Would not it be preferable to teach a mental construct that allows a person to frame a particular exposure in its proper context? Fresh fruits and vegetables are a key part to a healthy diet, and it does not matter if a study shows that broccoli or blueberries reduce cancer—you should be eating them (because they were part of a healthy diet, but don't forget to wash them to remove pesticide residues and infectious agents). If a person learned to look at the current health report and weigh it against their modifiable exposome, they may be more likely to place it in the proper context.

While many of the issues and concerns mentioned previously are addressed in the behavioral sciences and risk science, the general public is not so concerned with decision-making processes, biases, and probabilities. We know that most people are not intuitive risk scientists, but rather they are looking for an easy way to frame the complexity of the data with which they are bombarded. The exposome provides this framework and can introduce biological concepts in a context that makes it easier to remember over time. One does not need a college course in biology to understand the key components. When it comes to exposures that affect our health, the exposome stands out as a workable and teachable model that can be used to improve health. As part of the HERCULES Exposome Research Center at Emory University, the Community Advisory Board used the exposome concept to teach people in their neighborhoods about the environment. Michelle Kelger and Melanie Pearson worked closely with the community members to provide the materials and information needed to deliver this information to the communities. At Columbia University, the Columbia Center for Children's Environmental Health, investigators Frederica Perera, Julie Herbstman, and Virginia Rauh, and others have built a decades-long community partnership to address environmental exposures, and this work has led to marked improvements in health and even governmental policy changes. Community-based participatory research (CBPR) strives to involve the affected communities in all stages of research from design to data collection, and delivery of results. While CBPR is useful for university-based research projects, we are also seeing more exposome-based citizen science. Various monitors of pollution are available to the general public and many are using such systems to monitor air or water quality in their neighborhoods and using that data for personal decisions or to take to politicians to illustrate a community health concern. Simple programmable computers, such as the Raspberry Pi, can be coupled with sensors to create inexpensive and educational monitoring devices.

9.10 Obstacles and opportunities

The transfer of knowledge about the exposome should not be difficult from an educational standpoint. The primary challenge for the field is the general acceptance of the concept itself. Members of the general public are inclined to embrace a holistic view of their health, and the exposome concept fits within this view. However, adoption of the exposome within the halls of academia requires those in the field of environmental health sciences to see the value of a unifying framework that expands beyond their traditional intellectual boundaries. Within the field of environmental health sciences, many of the concepts of the exposome are already being addressed even if the term is not being used, or if the vision is not as broad. Incorporating the exposome, especially the big data–generating approaches, into courses within environmental health sciences is necessary. The explosion of complex datasets is not going to slow down anytime soon. In fact, trainees who ignore this unavoidable truism will be at a significant disadvantage. Integration of the exposome concept into medical education may be a different matter. The medical profession is very thoughtful and careful about alterations in their information-dense curriculum. It may be best to start with continuing education (CE). Developing CE courses that address certain aspects of the exposome could become a useful tool in medical education. However, improving environmental health science education among medical professionals is an important and attainable goal, and those with expertise in environmental health sciences should continue to explore ways of doing so. The field will need to prepare and deliver the appropriate type of materials for education and training for each of the aforementioned areas. The more resources that can be provided to the professor, scientist, or teacher willing to present this material, the more likely the effort will be successful. Ideally, courses that incorporate the exposome theme will become common in academia. With concerted efforts from scientists and public health professionals, the exposome can become a useful vehicle for teaching the importance of the environment in our health to those in other scientific disciplines and to the general public.

9.11 Discussion questions

1. Create a new version of the modifiable exposome with community on the left side of the *x*-axis and society on the right side. In essence, remove the role of the individual. Conversely, create a new version that plots individual on the left side of the *x*-axis and community or family on the right side. Discuss the differences and similarities among the different exposome plots.
2. Outside of class, explain the concept of the exposome to a scientific colleague and a nonscientific colleague. What was the response?
3. Plot your own exposome on Fig. 9.5. What are the activities, habits, and lifestyle choices that most greatly impact your health? Are these

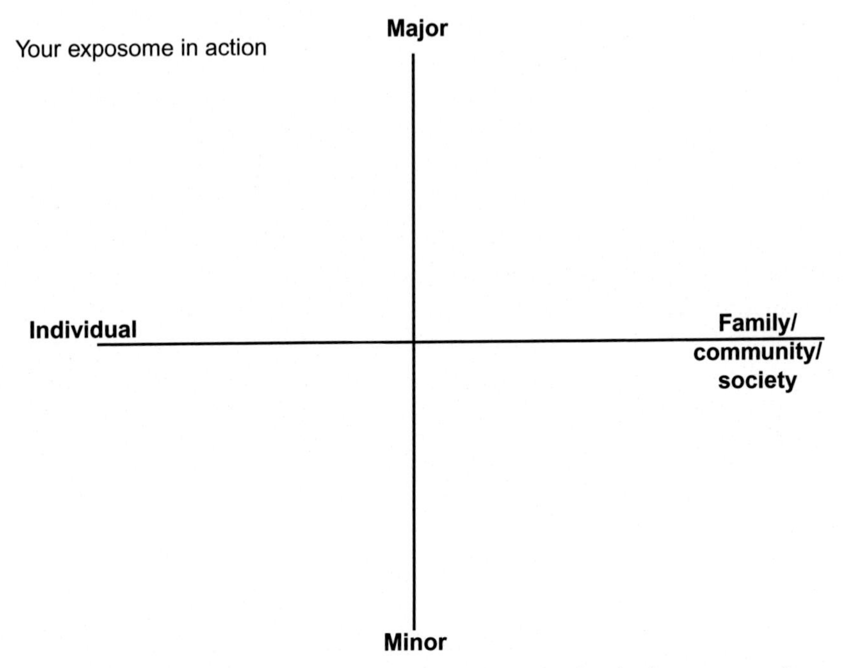

Figure 9.5 ***Your modifiable exposome.*** Use this graph to plot the key aspects of your life that influence your exposome. What changes can you make to drive these items toward the positive side of the graph? Which are the most important factors?

the areas that you think about the most? Set goals on how to shift these areas into a more positive direction.

References

Juarez, P. D., Hood, D. B., Rogers, G. L., et al. (2017). A novel approach to analyzing lung cancer mortality disparities: Using the exposome and a graph-theoretical toolchain. *Environmental Disease*, *2*(2), 33–44.

Khan, S. U., Khan, M. U., Riaz, H., Valavoor, S., Zhao, D., Vaughan, L., ... Michos, E. D. (2019). Effects of nutritional supplements and dietary interventions on cardiovascular outcomes. *Annals of Internal Medicine*, *171*, 190–198.

Offit, P. (2013). The vitamin myth: Why we think we need supplements. The Atlantic. July 19.

Sunstein, C. R. (2013). Simpler: The future of government. New York: Simon Schuster.

Thaler, R. H., & Sunstein, C. R. (2008). Nudge: Improving decisions about health, wealth, and happiness. London: Penguin Press.

Further reading

Ahmed, S. M., & Palermo, A. G. (2010). Community engagement in research: Frameworks for education and peer review. *American Journal of Public Health*, *100*(8), 1380–1387.

Burstein, J. M., & Levy, B. S. (1994). The teaching of occupational health in US medical schools: Little improvement in 9 years. *American Journal of Public Health*, *84*(5), 846–849, Super old reference (May 1994).

CHAPTER TEN

The exposome in the future

It is essentially immoral not to get the human exposome measured as fast as possible.

Gary W. Miller

10.1 Introduction

In yet another parallel to the genome, this chapter opens with a somewhat tongue-in-cheek quote from yours truly that I first uttered at the inaugural New York City Exposome Symposium in November 2018. The event was held at the New York Academy of Medicine and hosted by the Institute of Exposomic Science at the Icahn School of Medicine at Mount Sinai. Decades earlier, James Watson admonished the world with the phrase "It is essentially immoral not to get the human genome done as fast as possible" (*New York Times*, Natalie Angier, June 5, 1990). I do not agree that sequencing the human genome was a case of morality, although it was an exceedingly important scientific endeavor. However, I do believe that the exposome is just as important as the genome so I happily paraphrase the Watson morality clause for the exposome. My initial plan was to reuse the quote from Edwin Land from the opening chapter, and since I believe it provides the better framing for this closing chapter than my appropriated and modified quote, I will pivot back to it. To remind the reader, Land's innovative spirit was captured in the quote "Don't undertake a project that is not manifestly important and nearly impossible."

This chapter focuses on a vision for exposome research, and I can think of no better sentiment than Land's visionary statement. I was not aware of his quote when I started working on the exposome almost 10 years ago, but I am struck by how well it captures the ethos of the exposome. The exposome—that is, a concept intended to capture the complex and dynamic environmental forces impacting our health—is undoubtedly

The Exposome.
DOI: https://doi.org/10.1016/B978-0-12-814079-6.00010-9

manifestly important. Moreover, at times, it has appeared to be nearly impossible. Indeed, in 2005 when Wild coined the term, in 2010 when the National Academy of Sciences hosted the workshop on the exposome, and again when we starting drafting the proposals to develop research centers on the exposome,[1] it must have appeared nearly impossible to those involved. Yet the power in Land's phrase is the juxtaposition of the "manifestly important" and the "nearly impossible." As noted in Chapter 6, Innovation and the Exposome, innovation is all about pursuing that which appears to be nearly impossible, but it is the manifest importance that drives one to look beyond the impossible and come up with unexpected and often unpredictable solutions. What exactly did Land mean by manifestly? Some synonyms of manifest include evident, apparent, explicit, undisguised, unmistakable, unquestionable, and undeniable. I believe that obtaining a deep understanding of the environment's role in our health is important at a level equal to all of those descriptors. This also explains my impatience with colleagues who say that what we are trying to do is impossible. First, the use of the word totality[2] in the definition of the exposome does seem to make it impossible from a practical (and even intellectual) perspective, which is why Jones and I provided a modified definition (Miller & Jones, 2014). If you accept our operational definition, it is within the realm of the possible because it is a definable entity (the cumulative measure of environmental exposures and corresponding biological response). That said, the scale of what we are trying to do does put the exposome on the precarious boundary between impossible and possible, but that is exactly where one wants to be to incite innovation. As more people have been recruited to address the challenges surrounding the exposome, they have exerted more effort and applied more creativity, which gets us a little closer to possible. At times the improvements may seem incremental, but I predict we will see dramatic and saltatory leaps in technology that will soon make the human exposome seem more possible than impossible (and remain manifestly important). This chapter embraces that view and proposes some of the directions and avenues that can help transition the exposome from the quirky concept to scientific reality.

[1] I am assuming the leaders of HELIX, HEALS, and EXPOSOMICS felt similarly to those of us who were crafting HERCULES.

[2] Early papers state the term "exposome" refers to the totality of environmental exposures from conception onward, and has been proposed to be a critical entity for disease etiology (Rappaport & Smith, 2010; Wild, 2005).

Whenever I struggle with the magnitude of challenges incumbent upon exposome investigators, I often look for similar initiatives that are pursuing something similar in scale and ambition. Essentially, I am attempting to calibrate my own and the field's level of over-ambitiousness with the rest of science. The efforts to build the Large Hadron Collider to identify subatomic particles, such as the Higgs boson, and other physics projects provide great examples of extraordinary projects. These efforts are critical to our understanding about the nature of the world, but they do not have an obvious impact on our state of humanity. I recently learned about the Earth Biogenome Project and was comforted by the sense of reasonableness of the exposome approach in relative comparison. The Earth Biogenome Project has a "simple" objective to sequence, catalog, and characterize the genomes of all of Earth's eukaryotic biodiversity in a 10-year period. Let that sink in. There are more thanr 1.5 million eukaryotic species on Earth, spread among 9000 taxonomic families. In the same amount of time that it took to sequence the human genome, this organization plans to sequence 1.5 million species. Yes, it is extraordinary, but is it unexpected? If we go back to Chapter 2, Genes, Genomes, and Genomic: Advances and Limitations, this approach follows the familiar gene-centric pattern. This is not an indictment of the approach. The sequence information will create an amazing catalog of biology. It will provide a molecular and computational index of all of the collected biospecimens found in museums and laboratories around the world. It will provide a mineable data resource that will advance our knowledge of the species that populate the Earth. Geneticists know how to work at scale. This is a logical step in the advancement of their field.

I often refer back to the article "The Human Genome Project: lessons from large-scale biology" by Collins, Morgan, and Patrinos (which opened with the Watson immorality quote; Collins, 2003), and I suspect that the leaders of the Earth Biogenome Project have studied it carefully. The Earth Biogenome team provides clear objectives; they have mapped out a plan with clear milestones and have assembled an international and collaborative team. I applaud the leaders of the Earth Biogenome Project for their audacity and their precision in execution. There are lessons here to be learned. At the same time, I harken back to the arguments provided earlier in this book—genes do not tell the whole story. They do tell the whole story of the genetic makeup of the Earth's Biogenome, but not its biology. To do that we must understand the environmental forces that act upon the genes.

10.2 Features of a human exposome project

To provide a systematic analysis of the environmental forces impacting health, the exposome must be transformative. As noted in the opening chapter, I do not take lightly using the term paradigm for the exposome (and am still apprehensive seeing it in the title of this book). I am not convinced that the theory and methods behind the exposome concept have achieved this stature yet, but I do believe that the current conditions are indicative of the state that directly precedes a paradigmatic shift within a particular discipline. Specifically, as noted earlier, fields approaching paradigmatic shifts often find themselves in states of crisis (Kuhn, 1962). There is inner turmoil in a field because existing approaches are inadequate for the current or emerging questions. Our inability to address the health effects of exposures that occur as mixtures-that is, essentially all of them-is a case in point.

Over the past years, I have often been asked, "What would a human exposome project look like? How would one approach measuring the exposome?" I typically defer answering this type of question because of the sheer complexity of the problem and an inability to explain a massive project in just a few sentences. It is good that the questions are being asked. If people are interested in how such a project would be pursued, it is likely that they view it as an area worthy of inquiry. Here, I will attempt to answer these questions.

10.3 Measuring the exposome

In a manner analogous to the Human Genome Project, a well-coordinated international collaboration will need to be established to begin a large-scale study to measure the human exposome, although efforts already underway are starting to make progress on certain aspects of the exposome. This chapter outlines a possible strategy for moving a human exposome project forward, including the importance of data-sharing and international collaboration. I am a fan of the word collaboration, but as noted earlier not so much of the word consensus. Consensus means broad unanimity, as in unanimous, as in everyone agrees. If everyone agrees on how to approach the exposome, then we are doing it incorrectly. There will be and should be significant disagreement and debate about how to do this. Consensus is also often used in discussion of professional opinion. The exposome should not dabble in opinion. The exposome must be data-driven and verifiable. Consensus is also often used to suggest that research is no longer needed, that the scientific overlords have spoken. This is a very dangerous position in which to be. As scientists, we should welcome thoughtful criticism and a continued degree of skepticism. It is not easy to develop cogent arguments and valid scientific methods. The continued consideration of opposing views and approaches helps sharpen scientific thinking. The key is to create an inclusive process that accepts the input of representatives from multiple organizations but ultimately comes up with a process that is acceptable to at least a majority of participants and scientifically sound to an impartial observer (or an external and independent advisory group). Objections should be noted and considered throughout the process. Alternative approaches can be pursued and the results compared to the majority view.

A single author proposing a plan for how to pursue the exposome is the antithesis of the aforementioned plan. However, for the purpose of discussion and in the hopes of spurring lively debate, this is precisely what I will do here. I have interacted with and listened to hundreds of scientists who are interested in or are actively studying the exposome. Thus, my views have been shaped by numerous creative, thoughtful, and analytical scientists. I hope that I have learned from them and that my opinions have taken their expertise into mind. First, exposome research should involve a series of coordinated and parallel projects run from different sites throughout the world. It should not be one single massive project. We know from discussions of the World Wide Web that distributed networks are much more robust. Science works the same way. When the vision, execution, and administration of a project are not well aligned, large efforts can fail. For example, the National Children's Study was halted in the U.S.; in part because it was crushed under its own administrative bloat and lack of clear direction.

The European Commission has invested heavily in exposome research, and the EU2020 Human Exposome Network, which is funding nine projects, represents the largest exposome effort to date. It will be interesting to see how much coordination, harmonization, and collaboration can come from the individual projects (which will be discussed more later). I suspect that over the next few years, we will identify the scientific hurdles that require a worldwide initiative. Whether it is an effort parallel to the

Human Genome Project or a set of standards and a shared data resource to crowdsource the exposome is not yet known. Many believe that the Human Genome Project was an individual project, but in fact, it was a distributed effort across many organizations. There was a singular goal, but not a singular project from an organizational chart perspective. There were many steps that were performed at the same time across multiple institutions and a similar approach should be taken for the exposome. As shown in Fig. 10.1, we are at the stage of developing and testing enabling technologies. We are currently not in a position to pursue a human exposome project that includes 50,000 to 100,000 people, which is the level I estimate that will be needed to make causal associations for several major diseases.

The current exposome-related projects (primarily those in Europe and the U.S.) are building the infrastructure, testing out new methodologies, and determining if the improved resolution of exposures is adding value to the study of human health and disease. These projects and activities are occurring in parallel, but there has been little interaction among these projects and investigators. The European Commission is in a superb position to drive interaction among the new projects it is funding under the Human Exposome Network (more later), but there is still a major need to improve the analytical capabilities and the level of harmonization among the projects. Further, there has been insufficient effort to create a global consortium that will likely be needed to understand the diversity of exposures throughout the world.

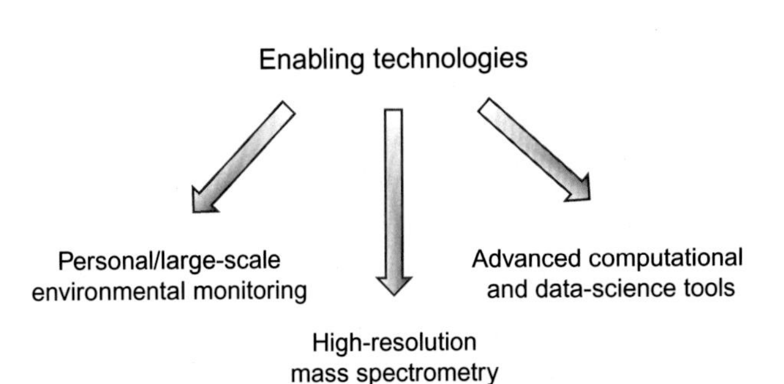

Figure 10.1 ***Enabling technologies.*** Exposome research requires ongoing development and refinement of methods and approaches to improve our ability to identify exposures. Continued development of personal environmental monitoring, high-resolution mass spectrometry, and computational platforms will help provide the field with the necessary tools to move forward.

Exposome-related research and a human exposome project are two distinct activities. The former is underway, while the latter is a mere goal at this time. The work currently being conducted is providing the foundation for an eventual human exposome project (or a virtual project that can be assembled retrospectively). Thus, one could divide exposome research into two divisions. The first is the work that is conducted up to the point that a true human exposome project is initiated. The second is the human exposome project itself. There are numerous exposome-related activities underway, some of which are described later. The idea would be to use the data and tools being generated by these ongoing projects to design and conduct a full-scale human exposome project.

Increased communication among the various investigators should occur, but during this build-up phase, nothing should be dictated. The projects funded by the European Commission should and will received guidance from their funders, but ideally innovative approaches are encouraged even if they diverge from the original plans. The various groups should conduct their work in the manner that best advances the science. In essence, all of the ongoing research will serve a series of large-scale pilot studies to determine the best tools and approaches to be used in more comprehensive projects in the future. However, the teams that are conducting exposome-related research should start meeting on a regular basis to exchange data and insight. The stronger the collaborative relationships are over the coming years, the higher the likelihood that the groups will be willing to enter into a multisite project to conduct the large-scale studies that will be necessary. The best way to do this is not clear, but societies and journals focused on the exposome would be a step in the right direction. The field needs to establish the constellation of beliefs and values to help guide efforts over the upcoming decades.

10.4 What is in us? What is near us? Does it matter?

At this point, it will be helpful to review what it is we are trying to get out of the exposome. What should be the focus? Obviously, the exposome is going to be heavily focused on the chemicals that reside or did reside in our bodies. We want to know what chemicals should be in us and what chemicals should not be in us. We want to know how the chemicals alter our biology. We want to know how exposure to a class of chemicals impacts our endocrine system, modifies our DNA, or alters synaptic signaling. Yet we must also consider the nonchemical exposures whether they be physical or social. The physical forces of sunlight and artificial light are major regulators of our physiology. Radiation from space, nuclear energy sources, the earth, and the machinery around us are also known to impact health. Noise from industries, traffic, music, and occupational work can cause hearing loss and systemic stress. What is the effect of our communication systems? Electromagnetic waves? Then, of course, the sociological issues that include the effects of trauma or violence, poverty, and oppression are also known to be major drivers of health and disease. Cataloging and measuring these exposures may be more difficult, but as I have emphasized, since we know that these forces are having an impact on health there must be biological evidence of their effects. Therefore, studies that examine the biological indicators of the social determinants of health should be emphasized and supported by health-related funding agencies. In the U.S., NIH spends the majority of its funding on molecular and biochemical studies and not enough on the social determinants of health. An exposome approach that links the social forces to the downstream biological effects may be an appropriate fit in the NIH model. As noted previously, telomere length, epigenetic patterns, protein modifications, and immunophenotyping can all be affected by psychological stress or trauma and could contribute to an exposome risk score (ERS).

The field is developing and testing sophisticated methods for measuring the biological soup that is represented by our blood, and these approaches will greatly enhance our understanding of how these chemicals are affecting our biology. We also want to know what is near us. In a systematic and parallel fashion, we must measure the complex exposures outside of the body (what Lioy and the NRC report referred to as the eco-exposome, but that I strongly believe to be part of the exposome). Where are these chemicals coming from? Does leaving closer to a factory actually increase the deposition of emitted chemicals to a person? What if they drive past the factory every day? Does it matter if the exposure occurred in combination with another pollutant or nutrient? Are there common patterns of exposure? Can we model exposures using some of the same network approaches we are using for the metabolic networks in our bodies? Lack of information about the source of the chemicals limits the value of knowledge of the chemicals in our bodies and their biological effects. This is why the spatial analysis and mapping discussed in Chapter 8, Data

Science and the Exposome, are important. We are going to generate multiple complex layers of exposome-related data. We will be generating data across various domains, but that data must be used to inform the other domains. We will be dealing with data from molecular, cellular, animal, human, population, and environmental sources. We must work on integrating these complementary findings in ways that improve our overall understanding of how environmental factors alter our bodies. Investigators must be willing to work among the various domains. The data will likely be structured in a multilayer network, and this will require the use of sophisticated computational approaches as addressed in Chapter 7, Pathways and Networks, and Chapter 8, Data Science and the Exposome.

Does it matter? The question of whether or not the chemicals that are present in our environment and in our bodies actually harm our health is what the exposome is all about. The mere presence of a chemical in our bodies does not mean that there is an adverse effect. That said, when we test one chemical at a time in a toxicological setting, we can only conclude toxicity thresholds for that individual chemical in isolation. When combined with a thousand other chemicals, that low concentration may indeed exert adverse effects. As noted in Wild's original paper, epidemiologists need a higher quality of data to make such determinations (Wild, 2005). The exposome should be able to provide this higher fidelity data, but the disease association studies will still need to be conducted. It may be that we need a few more years of technological development before we have the type of data to achieve Wild's vision.

10.5 A basis in biology

The exposome should be viewed as a biological challenge. We want to know how the environment influences our biology. We must build upon our current understanding of biological systems. Biochemical pathways, physiological systems, and mechanisms of toxic action must be the foundation of the exposome. There must be biological plausibility at every step (plausibility allows a level of biological uncertainty to still be factored in). The molecular pathway/network model is an excellent starting point, in that much of the key information is already available. Basing a model upon the known biological pathways and networks grounds the model in an already characterized system. For example, the Kyoto

Encyclopedia of Genes and Genomes (KEGG) pathways could serve as a starting point (Kanehisa, Goto, Sato, Furumichi, & Tanabe, 2012). Unfortunately, the future of KEGG is not clear, and it has not kept pace with the rapid advances in scientific knowledge. Hopefully, KEGG will find additional financial support to continue. Some of the crowdsourced efforts, such as WikiPathways, may need to step up to fill the emerging void. It could use biological pathways that have already been identified to be perturbed by environmental insults, such as key endocrine pathways that can estimate biological outputs from alterations in estrogen and testosterone receptor binding and activation. Structure activity/toxicophores could be built to allow for multiple compounds to be screened in silico. Computational models from multiple pathways can be assembled together into more sophisticated models as discussed in Chapter 7, Pathways and Networks, and Chapter 8, Data Science and the Exposome. Although the Tox21 initiative is built upon molecular and cellular pathways, the motivation for it comes from more of a regulatory perspective. While the exposome could have a major impact on regulatory decision-making, such issues should not be the basis for conducting the research. If the exposome is perceived as a mechanism to increase regulation, it will face strong resistance from various groups that must be involved in the long-term solutions. We should be evaluating the lifespan of the chemicals that are doing us harm. What are the needs and uses of the goods that contain or require the chemicals? Are there alternate manufacturing processes that can reduce emissions and exposures (Zimmerman, Anastas, Erythropel, & Leitner, 2020)? If we want to optimize health and minimize disease and suffering, academics must involve industry and manufacturing partners in the solutions. While many questioned the approach and cost of the Human Genome Project, few questioned its scientific value, and it was supported by both sides of the political spectrum. The exposome should be approached in a similar manner. The exposome will be providing insight on the nongenetic factors involved in health and disease. We need this information. The focus should be on the acquisition of knowledge.

10.6 What are the exposome deliverables?

There are several potential outcomes of exposome research. The first major deliverable will be improved methods for measuring and cataloging the chemicals (exogenous and endogenous) in our body. This will likely be the result of improved mass spectrometric technologies and other techniques that provide exquisite separation and identification in biological samples as detailed in Chapter 5, Measuring Exposures and their Impacts: Practical and Analytical. Improved identification of unknowns will be essential to convey confidence in the data that catalogue the chemicals in us. The next major deliverable will be validated assessments of the impact of those exposures on our biology (that is, the associated biological responses in the revised definition). This will likely come from a combination of approaches including some of the newer methods to assess DNA modifications at a genome-wide scale, proteomics, and other advanced methods such as immunophenotyping. Enhanced exposure assessment from improved technology benefits from being collected at the same time that data from the internal measures are being collected. In essence, by being yoked to the biological measures, the data from the exposure assessment will yield a new dimension of insight. Another key outcome will be a mechanistically based computational platform(s) that enhances our understanding of how suspected chemicals (thousands) contribute to disease. Much of this may be built upon systems biology-based models of human diseases being generated by other fields, as discussed in Chapter 8, Data Science and the Exposome. One of the most important deliverables will be a suite of tests or measures (an exposome test or index, a battery of tests, or perhaps an ERS—see later) that can be used in population studies to examine the relationship between complex exposures and human disease and in clinical settings to improve healthcare.

This book was written from an academic perspective as an intellectual exercise to better understand the exposome. The goal is to understand how our surroundings influence our health from a biological and scientific standpoint. But we do live in a practical world with a need for practical solutions and decisions. Regulatory agencies in the U.S., Europe, and other parts of the world must make decisions about environmental factors. How can these groups use the exposome? One of the challenges of regulatory agencies is that are in that tricky business of prediction. They are trying to predict which factors are most likely to be contributing to human disease and then predicting which interventions will best mitigate these risks. Many of these agencies are looking to exposome-type projects to guide them in their predictions, but this is a bit premature. That said, these same agencies have collected massive amounts of data that are very relevant to the exposome, and they are proceeding with projects to develop predictive algorithms, but we do not know if the data we are collecting are valid and if the projections and predictions are scalable. Can our current models adapt the complex data being generated? Can agent-based models handle exposome complexity? The science described in this text should help inform these agencies, but until the approaches are empirically tested and large-scale population studies are conducted, it may be unwise to integrate them into regulatory decisions.

10.7 Exposome risk scores to complement polygenic risk scores

A major goal for the exposome should be for its concept and information to be integrated into the care of the patient. This may be one of the harder goals to achieve, but it must remain a goal. As noted previously, it would be preferable to focus on assessing current abilities to respond to insults and not on predicting future outcomes. In essence, how robust is an individual based on the current assessment of his or her exposome? Twenty years ago, the term allostasis was introduced by McEwen and Stellar as a way to describe the wear and tear that occurs in the body in response to repeated stress. Allostatic load has been used in some fields, but it has not gained widespread use. Some of the concern is that the measurements are somewhat crude, though readily measurable [e.g., systolic and diastolic blood pressure, waist–hip ratio, serum HDL and total cholesterol, plasma levels of glycosylated hemoglobin, serum dihydroepiandrosterone sulfate (DHEA-S), overnight urinary excretion levels of cortisol, norepinephrine, and epinephrine]. While these measures can provide an assessment of the hypothalamic-pituitary-adrenal axis and responses to stress, these type of data do not provide the "associated biological response" in our definition of the exposome. For this, we must measure some biological changes that persist over time, such as epigenetic changes, DNA adduct formation, telomere length, and other aspects of cellular damage. Approaches that measure the body burden of environmental chemicals can provide an index of exposure, and when compared to the relative "associated biological response," one can assess the relative ability of an individual to respond to subsequent insults. The efforts to define and measure biological aging are analogous and may well fit into the exposome paradigm. The exposome field should work closely with aging biologists to explore such collaborations. Thus, for the exposome we need to ascertain the most appropriate battery of tests to measure this associated (and cumulative) biological response that would be analogous to the allostatic load. Combined with a comprehensive assessment of exposures, such a measure could become part of an exposome index that could ultimately be used in a clinical setting.

I sketched out an exposome index (similar to the ERS) in the first edition of this book, but did not work to advance the concept until very recently. For the purpose of discussion, Fig. 10.2 shows some possible components of an ERS similar to what my colleagues and I proposed in a recent publication (Vermeulen, Schymanski, Barabási, & Miller, 2020). Briefly, an ERS could help summarize the nongenetic risk factors for disease. This would help identify types of exposures that pose the greatest risks to human health. Once again, building upon the genomic framework, polygenic risk scores provide an important example. Genetics has the advantage of being static, so that one can model the combined effects of multiple genetic polymorphisms. That said, the genetics field has struggled with the development and application of the polygenic risk score due to statistical issues and interpretation of the results. The exposome field should follow these developments carefully and not try to promise too much for such a score. It is likely that an ERS would have disease- and exposure-specific components (see Fig. 10.2). However, once exposome data are collected at scale, it should be feasible to develop an ERS framework that can be used in conjunction with the polygenic risk score

Exposome risk score (ERS) components

Body burden of exogenous chemicals organized by chemical class and biological targets

Assessment of endogenous biochemistry, such as nutrient profile, microbiome variability, enzyme activity variants

Behavioral factors—physical activity, location, social factors, daily stressors

Aspects of biological response, such as inflammation, allostatic load, clinical measures, DNA methylation, DNA adducts, telomere integrity

Figure 10.2 ***Exposome risk score.*** As a means of stimulating discussion a list of possible components to an exposome index is shown. The idea is that each of these components could be measured in an individual at a specific point in time. Future studies would need to empirically determine whether or not these values correspond to changes in health status, which would allow refinement and retesting.

essentially bringing the GWAS (genome-wide association studies) × EWAS (environmental-wide association study) from a population level down to the individual level.

Fig. 10.2 shows some of the tests that could be run to assess exposures and exposome-related biological effects. Our DNA stays with us for life. While DNA repair mechanisms are continually working to maintain the integrity of our genetic code, there are lasting alterations to our DNA in the form of epigenetic modifications, adducts, or outright damage (deletions, cross-linking, etc.). Telomere integrity is considered to be one way to assess the health of our DNA. Thus measurement of several markers of DNA integrity provides a measure of our lifelong genetic insults and our ability to respond to said insults. What is measured on a particular day is the summation of these injuries, repairs, and regulatory responses. Our exposure to chemicals, and the absorption, metabolism, distribution, and excretion of them, is one of the most important aspects of the exposome. When we think about how to reduce our exposures to certain chemicals, we need to know which chemicals end up in our bodies relative to the presumed exposures (from use of air, soil, or food sampling data). The collection of chemicals in our bodies is the result of not only our own metabolic processes, but that of our microbiome. Is it necessary, though, to measure every possible chemical? Given that there are nearly 100,000 synthetic chemicals in our environment, this would seem to be an impossible and unnecessary task. It would be better to identify a set of chemicals or chemical classes that are representative of our complex exposures, but this would need to be data-driven with a priori determination. The high-resolution mass spectrometry approaches described in Chapter 5, Measuring Exposures and their Impacts: Practical and Analytical, coupled with data dimension reduction, may be a critical step in cataloging the exposome. Some examples of how an ERS could be constructed are shown in Fig. 10.3.

While an ERS addresses many of the body's responses to exposures, it does not capture the source of many of the external influences. For example, physical activity can be addressed by personal accelerometers in smartphones or similar devices. However, gleaning information on social determinants of health will be more challenging. The impact of stressful life experiences can be captured as part of the assessment of epigenetics, but determining the occurrence of such a life stress is more challenging and will require input from the subject. One of the interesting aspects of an exposome index is the ability to test its validity with tools that are

Disease-specific	Exposure-specific
Exposure Profiles A, B, C, D	*Exposure Profiles E, F, G, H*
A increases risk by 12%	E increases risk of diseases I, J, K
B increases risk by 20%	F increases risk of diseases L, M, N
C increases risk by 8%	G increases risk of diseases O, P, Q
D decreases risk by 35%	H decreases risk of diseases K, M, Q

Figure 10.3 ***Exposome risk score (ERS) models.*** An ERS could be disease-specific or exposure-specific. For a given disease or condition (for example, asthma) there would be exposure profiles (not individual chemicals, but networked or combined exposures) that would increase disease incidence or exacerbate symptoms. The exposure profiles may be additive (this would be gleaned from the large-scale exposome studies) or competitive. Assuming the effects are additive, a person with A, B, and C exposome profies would have a 40% increase risk of disease (the sum of the 12%, 20%, and 8%). Each profile would be data-driven and disease-specific and could be ranked by decile or the most appropriate divisions as determined by the data. For an exposure-specific ERS a person's decile rank for that particular profile would predict an increased risk for each disease. For example, the top decile of profile E may be associated with an increase in the risk of cardiovascular disease (I) by 15%, diabetes (J) by 50%, and eczema (K) by 60%. The ERS would be a probabilistic model that indicates that the patient is at increased risk because of the exposures. Ideally, ERSs would be introduced alongside recommendations for reducing that exposure profile. *ERS*, Exposome risk score.

already at our disposal. Prospective epidemiological studies could be designed to test the predictive validity of certain components of the ERS, as could longitudinal clinical research trials. It would be best to focus on health outcomes that could be measured over the course of few years. My colleagues at the Mayo Clinic recently proposed a framework for incorporating multiomic data into clinical research (Fig. 10.4). The ability to fuse data from these multiple streams remains an analytical hurdle for the exposome (and biomedical science, in general).

10.8 Setting priorities

One of the most daunting challenges of exposome research is not technological. It is philosophical, logical, and practical. Since we are trying to assess such an incredible breadth of factors, where does one start? If a person or group is designing an exposome study, what are the first steps? What I have observed is that many groups are trying to do too much too

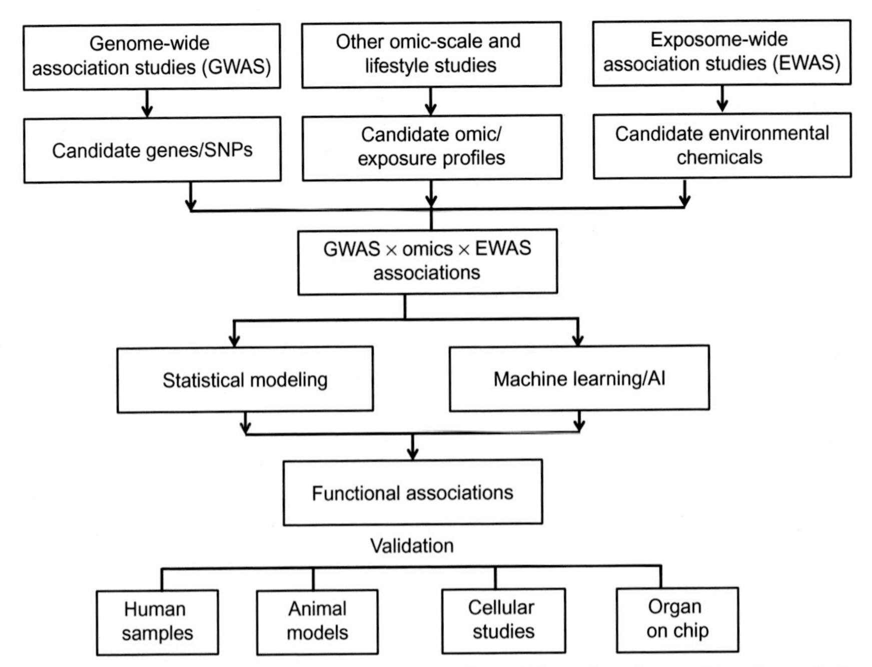

Figure 10.4 ***Omic integration.*** The potential workflow for the multiomic analysis needed for exposome research. Data is being generated from multiple sources. This includes genomics and proteomics. Exposomics data must be merged with these other streams of data. It is likely that descriptive, supervised, and unsupervised approaches will be needed. *Modified from Cheung, A. C., Walker, D. I., Juran, B. D., Miller, G. W., Lazaridis, K. N. (2020). Studying the exposome to understand the environmental determinants of complex liver diseases. Hepatology, 71(1), 352–362.*

fast. Initiating a study to capture as much as one can measure is likely going to provide less the optimal data for hundreds of outcomes. I believe that a better approach is to provide high-quality exposome-level data for specific diseases or specific outcomes. The focus should be on the measures for which we have reliable and state-of-the-art methods, but to collect and biobank samples that can be analyzed and reanalyzed at later times with the anticipation of improved technologies. In a complementary fashion, one can look at a focused set of exposures on multiple health outcomes. Exposome studies should rigorously test the limitations as they move toward the convergence of multiple comparisons from hundreds of exposures and hundreds of outcomes. This is an example of where staging can be so critical. What are the stages that suggest that the approach is working?

One reality is that some laboratories are better than others for nearly every assay. A system that supports the best facilities to generate the data in a harmonized fashion will be superior to one that allows labs with little experience to contribute lower quality samples. Alternatively, creation and optimization of an analytical system that is fit for purpose and replicable may facilitate multiple sites having harmonized capacity. The Human Genome Project made sure to have very high-quality standards for the groups that were contributing sequence data. An initial step may be creating a central repository for the human exposome that sets clear standards for what is needed for deposition. Then funding agencies could require grantees to conduct studies that adhered to that standard. There are several challenges to this approach. Decisions that are needed for setting up a central repository can be contentious. Who will be in charge? How does one balance scientific expertise and academic freedom with structure and bureaucracy? Current repositories for metabolomics data are insufficient for exposome-based needs and novel annotation approaches since they were not designed for environmental chemicals and their metabolites that do not have standards sitting on the shelf.

10.9 A human exposome project?

Six years ago, I asked if was it too early to start discussing a human exposome project. It was not clear at that time if we were ready for that Santa Fe Meeting to launch such an initiative. While more proof of principle data must be collected, it is now time to start having this discussion. The time to provide the complementary environmental analog to the human genome project has arrived. Whether it achieves the level of being a biological index of nurture is not clear, but there is no doubt we need to focus more resources toward our environment. Over the next 10 years a human exposome project could be underway, but it will take a considerable amount of careful thought and planning. In 2020 the European Union launched multiple projects under their Human Exposome Network initiative (Disclosure—I receive funding from one of these projects and serve as an advisor on another).

As shown in Figs. 10.5 and 10.6, planning for the execution of a human exposome project is a long-term enterprise. One of the most exciting aspects of the outline shown is that so many of the activities are

Proposed staging for an international exposome initiative

1. Formation of an executive or steering committee
2. Establishment of monthly teleconferences or web conferences
3. Initial kick off or planning meeting(s)
4. Identification of key measures (exposome index/risk score)
5. Establishment of data sharing policies
6. Establishment of a data hosting site(s)
7. Recommendations for funding initiatives
8. Cultivation of research dollars (government, industry, academia, private)
9. Conduct large-scale collaborative studies

Figure 10.5 ***Proposed staging of a human exposome project.*** Pursuit of the exposome will require a highly coordinated effort with international partners. This figure outlines the key steps that could be followed in the development and pursuit of such an undertaking.

already occurring (as illustrated in Fig. 10.6 in the year 2020). For example, the enabling technologies are under development, the exposome is being discussed at annual scientific conferences across multiple disciplines, and there is more frequent exchange among the research groups across the world. The interest in exposome-related research is growing, resources are being allocated, and energy is being devoted to the idea. Momentum for the exposome is building. My department chair Andrea Baccarrelli recently tweeted[3] in January, 2020 that we are "entering the decade of the exposome." I agree, and the next 10 years should be transformative for the field. The goal should be to harness the results of these ongoing activities, provide a clear focus for ongoing research, and then work to direct future initiatives that allow us to pursue the most exciting and informative science.

10.10 Global exposome harmonization

At the time of this writing, several of my colleagues are working toward a global project focused on harmonizing some of the key methods that will be needed for exposome research. Such a global effort will bolster exposome research by rigorously testing the potential methodologies and demonstrating their reproducibility. Seemingly mundane items such

[3] Tweeted: having posted news or information on the social media platform Twitter.

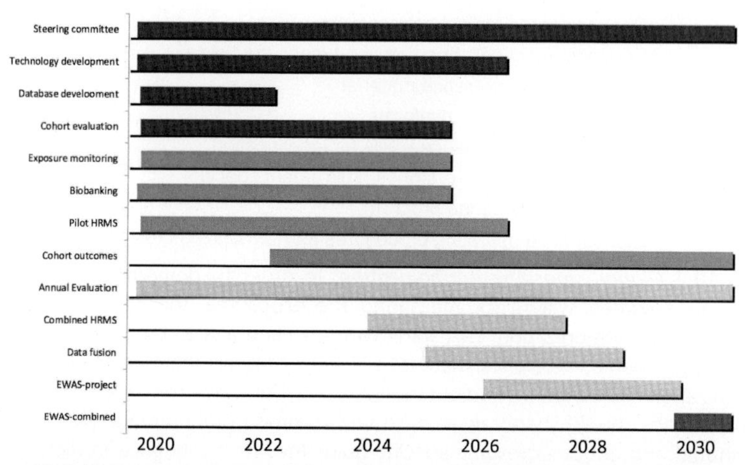

Figure 10.6 ***Human exposome project timeline illustration.*** Ten-year plan for the human exposome project. There are several exposome-related projects underway. Over the next few years, these unrelated projects will identify the best technologies and approaches. However, a human exposome project will require coordination among the various groups. The goal is for the groups to pursue the second round of studies in a manner that allows combining all of the data into a single exposome model or database. This would increase the statistical power, provide geographic and genetic diversity, and have would leverage the resources of all of the various research centers. *EWAS*, exposome-wide association study; *HRMS*, high-resolution mass spectrometry.

as ontologies (the vocabulary used to describe exposures, doses, conditions, health parameters, symptoms, etc.), standard operating procedures, reporting guidelines, reporting of confidence levels, and sensitivity analyses must be developed if we are going to be able to compare data across research sites (Figs. 10.7 and 10.8).

I believe that we need to form an interdisciplinary scientific society to help define key principles. I refer to Kuhn's definition of paradigm as the constellation of beliefs, values, and methods (Kuhn, 1962). The exposome does not have a clearing house for such information. There are several competing ideas, concepts, and approaches and much of this is due to the relative youth of the field. It is appropriate to have such disagreement as the science evolves; however, in order for an outsider to adopt an exposome-based approach, it is necessary for there to be a certain level of clarity. A society could provide the necessary intellectual home for these ideas. Development of an initial ERS could be a useful topic for an initial exposome meeting.

Commitment to radical transparency
Commitment to collaboration to advance the field
Shared pooled reference material
Shared bioinformatic platforms
Standardized confidence levels for identification
Validation among laboratories (instrument specific)
Investigator exchange program
Standardized operating procedures for harmonized projects

Figure 10.7 ***Global exposome harmonization.*** In order to demonstrate the coordinated and validated system for measuring the exposome, the field must develop approaches for harmonization. This starts with a willingness to be open and collaborative. Shared pooled reference material will help compare data across laboratories. Harmonization should be instrumentation-specific, but strive for concordance across platforms. Labs should share raw data to test common bioinformatic platforms. The field should also adopt exposome-specific guidelines for confidence in identification of features, which will likely differ from that used in general metabolomics. Also, laboratories should be willing to adopt standardized practices to facilitate large-scale multisite projects.

Below I outline a potential list of the values and beliefs that an exposome society could promote and I encourage colleagues to contribute to their further development and refinement.

1. To advance the science of the exposome.
2. To promote the concept of the exposome and the importance of our environment in health and disease.
3. To provide leadership for exposome research.
4. To enhance communication among exposome researchers across the world.
5. To enhance communication to scientists not involved in exposome or environmental health research.
6. To develop and provide access to innovative educational programs.
7. To increase funding for exposome-related research from government and private entities.

Besides working to develop an ERS, an important agenda item for an exposome society could be to simply define the exposome. This issue is raised on a regular basis and causes confusion in a variety of settings. I believe the definition that Jones and I have provided is easier to operationalize, but I am open to modifications (Miller & Jones, 2014). A scientific society structure could help address this "definition issue." As outlined in Chapter 1, The Exposome: Purpose, Definitions, and Scope, we must avoid Balkanization of the exposome and over derivatization of the word.

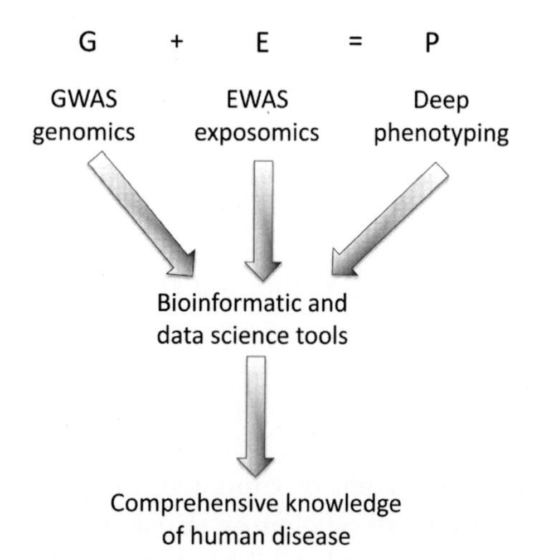

Figure 10.8 ***Convergence diagram.*** In the review by Vermeulen et al. (2020), we suggested how having exposome data at the scale and quality of GWAS would allow a true convergence of genetic and nongenetic contributors to disease. Combining exposome-wide data with genome-wide data and comprehensive phenotyping will allow scientists to determine the relative contributions of our genes and environment to health and disease. This knowledge can help drive decision- and policy-making that will improve personalized medicine and population health. *GWAS*, Genome-wide association studies.

We must present a unified and united front to those *outside* of the field. We must make the exposome relevant and clear to scientific colleagues across a wide range of disciplines, as well as making it comprehensible to legislators, government officials, and other decision makers.

The recent awarding of the international centers mentioned previously would seem to provide a superb foundation for planning a more expansive initiative. The formation of an international steering or advisory committee will be vital to the success of a futuristic plan. It should be composed of energetic and creative individuals who are willing to collaborate and cooperate. The key drivers of current exposome activities should be represented, as should the representatives of the funding organizations. There will be a need for frequent interactions in the form of teleconferences, list-services, and web-based updates. The convening of an initial kick-off meeting with key constituents will be critical. Getting the right people together to establish the vision and outline a plan is essential.

The steering body will need to address issues surrounding data-sharing and authorship policies. Data hosting sites and systems will need to be identified (and funded). Government science agencies will likely provide the majority of funding, but it will be very important to establish partnerships with industry to assist with the technological demands as well as working with private partners to raise funds for research and planning. Private support could help accelerate some of the initial brainstorming and planning.

A variety of studies and approaches will proceed over the next several years. The goal is to identify the best approaches being used by the various research projects and integrate them into the second phase of coordinated projects. For example, if a group has developed a novel personal environmental sensor, it would be ideal for all groups to employ such devices in the coordinated phase. Common approaches for identifying DNA adducts, performing metabolomics or exposomics, or the measurement of specific chemicals could be adopted. A core-based approach may be warranted in which many of these tests can be run at two separate sites for all of the human exposome project endeavors, with each site providing a particular expertise. In a similar fashion, investigators may want to use traditional survey instruments to determine how well these parallel the newer exposome-based approaches, but each site should be using the same instrument. To take such an approach, the various funding agencies would need to agree to the collaborative approach, which has precedence. It will be difficult getting the various scientists and teams to agree on common analytical workflows and approaches, but getting agreement on principles of data-sharing and transparency should be a priority. As noted earlier, there will be significant disagreement on what is the best way to address certain problems, but it will be critical to focus on the core problem of better understanding the environmental influences on our health. A large, well-designed, and rigorously executed study could have a dramatic influence on further studies in science. Thus, even after the end of the proposed 10-year plan there will be more to do. Ideally, the research conducted over those 10 years makes the world better appreciate how important the environment is to their health and helps put the environment in a more prominent place on the research agenda. In 10 years, we could have a new suite of tools and technologies that can be employed in other fields that have traditionally ignored the environment, have the preliminary data for examining environmental factors for diseases that have not considered environmental factors, and have permanently added the word exposome into the biomedical vernacular, and its concepts into healthcare models and into the minds of the general public.

The European Commission has made a major investment in exposome-related research as shown in Table 10.1. Japan has supported the comprehensive and nation-wide Japanese Environment and Children's Study. Unfortunately, a similar level of effort has not yet been seen in the U.S. The majority of health-related environmental research falls under the National Institute of Environmental Health Sciences (NIEHS), one of the 27 NIH institutes and centers. NIEHS receives a relatively small proportion of the NIH budget and has the disadvantage of being located over 200 miles away from the other NIH institutes. This places environmental health research at a considerable disadvantage in the U.S. A major effort to characterize the exposome could not be absorbed within the current NIEHS budget. It would require multiple NIH institutes working together to better understand the environmental drivers of disease. The distribution of research funds among the institutes is highly influenced by lobbying groups. It is difficult to criticize an organization that advocates for increased funding for a disease that has impacted the lives of their members, but this is one of the problems with research surrounding environmental influences of health. It is not a single disease or diagnosis. The impact of the environmental is strewn across multiple (nearly all) disease states. Allocating 2.5% of the NIH research to the institute focused on the environmental causation of disease is underwhelming to say the least. One may argue that other institutes do study environmental factors, but not to a degree that is concordant with the environmental contributors to disease, and they may lack the expertise in environmental health sciences that exists at NIEHS. Perhaps, the approach should not be to increase funding of the environmental health–based funding institutes, but rather increasing allocation to environmental health or exposome-based research at other research funding institutes. However, it is very difficult to argue for a research institute to expand its environmental health or exposome-based portfolios when the tools are not yet available. *For long-term impact, providing sensitive and validated tools and concepts that allow investigators outside the field to integrate the environment into their research should be a top priority.* While allocations across institutes may not change, if institutes recognize the potential utility of the exposome for their discipline and there is a framework for conducting such research, it is not ridiculous to think that research institutes could start encouraging work that incorporated the exposome.

Within the exposome framework, many research institutes are studying environmental factors. Studies that involve dietary influences, behavioral modifications, and social determinants of health, all factors outside

Table 10.1 The European Commission's Human Exposome Project Network awardees.

ATHLETE: Advancing tools for human early lifecourse exposome research and translation—Prof. Martine Vrijheid, Barcelona Institute for Global Health, Spain. This project will focus on measuring environmental exposures during pregnancy, childhood, and adolescence and link these early life exposures to cardiometabolic, respiratory, and mental health outcomes
EPHOR: Exposome project for health and occupational research—Dr. Anjoeka Pronk, Senior Scientist, Netherlands Organisation for Applied Scientific Research (TNO), The Netherlands. EPHOR will focus on the working–life exposome with a strong focus on exposures that occur in the workplace
EQUAL LIFE: Early Environmental quality and lifecourse mental health effects—Dr. Irene van Kamp, Senior Researcher, National Institute for Public Health and the Environment (RIVM), The Netherlands. EQUAL LIFE will focus on the impact of combined environmental exposures on child mental health and development across a range of ethnicities and social status
EXIMIOUS: Mapping exposure-induced immune effects: connecting the exposome and the immunome—Prof. Peter Hoet, Catholic University of Leuven, Belgium. This project is focused on immune-mediated diseases including asthma, allergy, and autoimmune disorders. It will combine exposome data with immune fingerprinting
EXPANSE: Exposome powered tools for healthy living in urban settings—Prof. Roel Vermeulen, Institute for Risk Assessment Sciences, Utrecht University, The Netherlands. This project has a major focus on the urban environment and access to exposome and health data for over 50 million Europeans. This will be combined with comprehensive high-resolution mass spectrometry-based exposomics on over 10,000 subjects
HEAP: Human Exposome Assessment Platform—Prof. Joakim Dillner, Karolinska Institute, Sweden. HEAP will develop a platform to integrate data from large population-based cohorts and analysis of the metagenome and epigenome, along with personal sensors for environmental monitoring
HEDIMED: Human exposomic determinants of immune-mediated diseases—Prof. Heikki Hyöty, University of Tampere, Finland. HEDIMED will focus on immune-related diseases such as type 1 diabetes, celiac disease, and asthma. The project will use advanced omic and data mining technologies to profile the exposome
LONGITOOLS: Dynamic longitudinal exposome trajectories in cardiovascular and metabolic noncommunicable diseases—Dr. Sylvain Sebert, University of Oulu, Finland. This project will focus on the lifecourse nature of the exposome in cardiovascular and metabolic disease with a goal of identifying optimal points of intervention
REMEDIA: Impact of exposome on the course of lung diseases—Dr. Sophie Lanone, Research Director, French National Institute of Health and Medical Research (INSERM), France. REMEDIA is focused on the impact of the exposome on the development of chronic obstructive pulmonary disease and cystic fibrosis using patient cohorts followed throughout life

Their projects started in 2020 with an investment of over 100 million euros. For more information see the website humanexposome.eu the traditional scope of environmental health science research institutes, are underway at other institutes. The more politically palatable solution may be to integrate more environmental health–related research into those research institutes and funding agencies that study diseases that are suspected of having a strong environmental component and to broaden the research to encompass the exposome. Organizations such as NIH need components focused on environmental factors, but the tools and approaches should be widely applied across institutes focused on particular diseases or conditions.

On the U.S. side, there simply has not been a coordinated effort to pursue the exposome. The Children's Health Exposure Analysis Resource (CHEAR) and Human Health Exposure Analysis Resources (HHEAR) projects attempted or are attempting to obtain more comprehensive environmental analysis on large human studies, but the vast majority of the work has been and will be conducted in a targeted fashion (measuring a specific number of known environmental agents, i.e., "looking under the lamppost"). It improved access to environmental analysis but falls short of the untargeted, unbiased, and expansive approach that my colleagues and I recently described as being necessary for the exposome (Vermeulen et al., 2020). The untargeted approaches capture much more chemical and biological information and provide greater opportunity for discovery. In full disclosure, I did lead one of the CHEAR laboratory hubs while I was at Emory University and am serving on the scientific advisory board to the HHEAR consortium.

In our 2020 perspective in Science (Vermeulen et al., 2020), we emphasized the importance of a human exposome project to be conducted at the scale of the Human Genome Project. Although I have mentioned some of the technical and design features, ideally such an undertaking would be led by an international group of experts. These experts must be able to influence funding agencies or be directly involved in the allocation of resources. Table 10.2 lists some of the characteristics I believe will be necessary for the success of a global human exposome project.

10.11 Obstacles and opportunities

Alas, the exposome faces many obstacles! As outlined in the above chapters, the exposome is complicated and will not readily reveal its secrets. However, we are making progress. The exposome provides

Table 10.2 Components of a global human exposome project.

1. The project must be run by an accomplished scientific management group (emphasis on accomplished scientists, but also with superb managerial skills)
2. Recruit people to the team who *want* to collaborate
3. The project must be driven by science not politics, but with the support of politicians
4. The project must have clearly articulated, staged, and quantifiable objectives
5. Public funds must be used judiciously with oversight from external groups like the U.S. National Academy of Science, the Wellcome Trust, other global scientific organizations
6. The proposed work must be exciting and well-resourced so that it attracts hundreds of top scientists

numerous opportunities and challenges for investigators interested in the influence of our environment on health and disease. Even if naysayers wanted to dispatch with the entire unifying notion of the exposome and jettison the word from the field, it would be difficult to do so. It is clear that pursuing a better understanding of the complex and lifelong exposures that impact our health is a worthy goal and the exposome encapsulates that goal. Environmental health sciences and exposome aficionados must be willing to address the issues surrounding the acquisition, analysis, and interpretation of big data, including the need to cultivate investigators who possess the necessary skills to perform the necessary studies. Without an aggressive campaign to recruit and develop these types of investigators, environmental health sciences will continue to play the role of follower within the biomedical community instead of the role of leader. Creative scientists with training in bioinformatics, systems and computational biology, and data science must be lured into the field. We must start developing the computational platforms that will help us start sorting and organizing the proverbial bricks that are already piling up but, more importantly, to organize the carefully constructed bricks yet to come. We should proceed with caution until we organize the brickyard.

Does the exposome represent a sufficient challenge to entice the next generation of scientists? From my perspective, the answer is a definitive "yes." Indeed, the vision for the exposome and its associated complexity can attract ambitious investigators. While the limitations of the gene-centric view of health and disease may have been intuitively obvious for those in environmental health science, they are becoming demonstrably so to the rest of the scientific community. The exposome provides the framework for the next major advance in the study of human health and disease. The exposome will also need specialists in the core sub-disciplines, but such investigators would be wise to expand their scientific repertoire to include computational approaches. Perhaps some of the energetic readers of this book will recognize the exciting opportunities provided by the exposome, start contemplating specific challenges, and eventually help define and refine this burgeoning concept.

If the exposome concept can help move the G × E needle in the direction of E by even a few percentage points, then the field can claim success. But until we are able to systematically identify the preventable causes of disease and develop exposome-level interventions that can improve human health on a population scale we will not be able to claim victory. Although genomics has the financial and technical advantage, being caged in public health, the exposome has its own distinct advantage. By its nature, exposome-level solutions are imminently scalable. Genetics may develop CRISPR-based methods for some diseases and treat thousands of people, but the exposome is poised to develop research programs, training opportunities, and policies that impact the lives of millions, if not billions, of people. The exposome is a simple concept with extraordinary implications.

The launching of the European Commission's Human Exposome Project Network in Brussels, Belgium in February 2020 may turn out to be the exposome's Santa Fe meeting (the defining event that will, in retrospect, signal the true birth of the field). Although it is European-centric, it represents the largest coordinated exposome effort to date. There are dozens of institutions with hundreds of investigators involved. However, there are several notable gaps. The first is that there is insufficient investment in technological advancement, harmonization, and data analytics. Without a concerted effort to harmonize data collection and analysis, it will be very difficult to compare data across the cohorts. My expectation is that these projects will accumulate a large amount of data on exposures, as well as on biobanked samples, for subsequent analysis. Development of a well-curated biobank that also has extensive exposure data would be of great value. If other countries can develop efforts of similar magnitude and partner with the European Commission, I believe such a foundation of coordinated studies with shared biobanked samples could be formed. Ultimately, we will need a combination of external exposure measures with a reliable, high-throughput, reproducible and omic-scale exposomic analysis that can be performed on >100,000 biobanked human samples.

The U.S. NIH has launched a study called All of Us. Its goal is to enroll over 1 million people in clinical research. Biosamples will be collected from each subject and stored at the Mayo Clinic Biobank (there will be 35 aliquots per subject). Although the patient recruitment is not done in a way that one would design a specific trial, it will nonetheless create a treasure trove of clinical data and biobanked samples covering the broad diversity of people in the U.S. It should be possible to create a virtual cohort that could enable a large exposome study of up to 100,000 subjects. Efforts are already underway to sequence the genomes of all participants (no surprise there), and exposome proponents must start positioning the field to compete for these valuable resources.

When Christopher Austin, Director of the NIH National Center for Advancing Translational Science (NCATS), spoke at our Emory Exposome Summer Course in 2016, he stated, "You can't wait for the technology...you have to move forward with what you have and the project will drive the technology." This was a major lesson from the Human Genome Project as described in Chapter 2, Genes, Genomes, and Genomics: Advances and Limitations, and Chapter 6, Innovation and the Exposome. The urgency of the Human Genome Project spurred technological innovation and corporate investment. As detailed in our recent manuscript, the technology needed for the exposome may be slightly ahead of where sequencing was when the Human Genome Project started (Vermeulen et al., 2020). Of course the exposome requires a compilation of technologies and data integration systems. Fortunately, these systems are rapidly advancing in ways that will support exposome-scale analyses for population studies and clinical medicine.

10.12 Conclusion

Again, I end here as I summarized the closing chapter of the first version of this book. The exposome is big science. This is not the type of project that can be conducted in an isolated laboratory. For the exposome to become a scientific and public health reality, it must involve investigators from across the world and from many disciplines. Such complex initiatives require sophisticated, yet nimble, organizational structures. A successful exposome initiative will require (1) involvement of academic, industry, private, and government organizations; (2) having a clear set of objectives; (3) advancement of technology across many platforms; (4) development guidelines for data storage, analysis, and sharing; and (5) mechanisms to assess progress and guide future steps. None of these steps are easy, but all will be essential. The technological and data advances will be complicated, but I suspect that the most difficult task will be coming up with a clear set of objectives. There are many factions participating in exposome research and all want to have their areas represented. I implore leaders to not try to please all of the constituents and distribute resources in a manner that leads to multiple mediocre efforts. Such an approach will dilute the efforts and energy and risk undermining the incredible progress that has already been made. Financial and intellectual resources should be focused on those areas that have the greatest potential to advance exposome research. As I alluded to earlier in the book, the exposome is an excellent foil to the genome. By having a similar level of resolution and quality of data, the exposome can reveal further insights into our genome. My colleagues ended our recent manuscript with the lines "together the combined knowledge of GWAS and EWAS would make it possible to systematically explore studies at the gene and environment interface, which is where nature meets nurture and chemistry meets biology" (Vermeulen et al., 2020). I believe that exposome research must strive to deliver the tools to do this. We need to provide data and knowledge that fits into the large framework of biomedical research. We must strive for the exposome to achieve the level of foil and "to contrast with and so emphasize the qualities of the genome (and vice versa)." When approached in such a collaborative manner with genome research, the exposome will unravel mysteries that are readily put in the context of our current knowledge of biology. As the title of my first manuscript on the exposome indicated, the exposome represents the nature of nurture (Miller & Jones, 2014), an ambitious, audacious, achievable, and manifestly important goal.

10.13 Discussion questions

1. If a human exposome project succeeded, how would it affect healthcare or public health? If certain chemicals were demonstrated to contribute to a particular disease, what strategies could be developed to mitigate these effects?

2. If you could have either full genome sequencing for full exposome analysis conducted on yourself, which would you prefer? Which one is more likely to lead to changes in your lifestyle or habits?
3. What modifications would you make to the exposome risk score?
4. Would the exposome concept be understandable to nonscience majors at colleges and universities? High school students? Middle school students? Elementary school students? Are students at these levels prepared to understand the complexity of the issue?
5. How can you introduce the concept of the exposome and its importance to your classmates, colleagues, family, and friends?

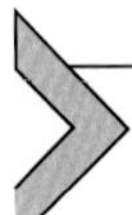

10.14 Additional resources

Human Exposome Project—Humanexposomeproject.com

European Commission Human Exposome Network—humanexposome.eu

U.S. National Institute of Environmental Health Sciences—www.niehs.nih.gov

U.S. Environmental Protection Agency Computational Toxicology/Tox21 initiative—http://epa.gov/ncct/Tox21/

References

Cheung, A. C., Walker, D. I., Juran, B. D., Miller, G. W., & Lazaridis, K. N. (2020). Studying the exposome to understand the environmental determinants of complex liver diseases. *Hepatology*, *71*(1), 352–362.

Collins, F. S., Morgan, M., & Patrinos, A. (2003). The Human Genome Project: Lessons from large-scale biology. *Science*, *300*, 286–290.

Kanehisa, M., Goto, S., Sato, Y., Furumichi, M., & Tanabe, M. (2012). KEGG for integration and interpretation of large-scale molecular datasets. *Nucleic Acids Research*, *40*, D109–D114.

Kuhn, T. S. (1962). *The structure of scientific revolutions*. Chicago, IL: University of Chicago Press.

Miller, G. W., & Jones, D. P. (2014). The nature of nurture: Refining the definition of the exposome. *Toxicological Sciences*, *137*, 1–2.

Rappaport, S. M., & Smith, M. T. (2010). Epidemiology. environment and disease risks. *Science*, *330*(6003), 460–461. Available from https://doi.org/10.1126/science.1192603.

Vermeulen, R., Schymanski, E. L., Barabási, A.-L., & Miller, G. W. (2020). The exposome and health: where chemistry meets biology. *Science*, *367*, 392–396.

Wild, C. P. (2005). Complementing the genome with an "exposome": the outstanding challenge of environmental exposure measurement in molecular epidemiology. *Cancer Epidemiology, Biomarkers & Prevention*, *14*(8), 1847–1850.

Zimmerman, J. B., Anastas, P. T., Erythropel, H. C., & Leitner, W. (2020). Designing for a green chemistry future. *Science*, *367*(6476), 397–400.

Further reading

Wild, C. P. (2012). The exposome: from concept to utility. *International Journal of Epidemiology*, *41*, 24–31.

Index

Note: Page numbers followed by "*f*," "*t*," and "*b*" refer to figures, tables, and boxes, respectively.

F

G

K

L

M

N

Printed in the United States
By Bookmasters